PORT CYBERSECURITY

PORT CYBERSECURITY

Securing Critical Information Infrastructures and Supply Chains

NINETA POLEMI
European Comission, Brussels, Belgium

ELSEVIER

Elsevier
Radarweg 29, PO Box 211, 1000 AE Amsterdam, Netherlands
The Boulevard, Langford Lane, Kidlington, Oxford OX5 1GB, United Kingdom
50 Hampshire Street, 5th Floor, Cambridge, MA 02139, United States

Library of Congress Cataloging-in-Publication Data
A catalog record for this book is available from the Library of Congress

British Library Cataloguing-in-Publication Data
A catalogue record for this book is available from the British Library

ISBN: 978-0-12-811818-4

For information on all Elsevier Publishing visit our website at
https://www.elsevier.com/books-and-journals

Publishing Director: Joe Hayton
Acquisition Editor: Tom Stover
Editorial Project Manager: Andrae Akeh
Production Project Manager: Punithavathy Govindaradjane
Cover Designer: Mark Rogers

Typeset by TNQ Books and Journals

CONTENTS

List of Figures vii
List of Tables ix
Acknowledgments xi
General Security Glossary xiii
Maritime Glossary xv
Executive Summary xix

1. Introduction **1**

2. Ports' Critical Infrastructures **7**
Maritime Environment: The Role of Commercial Ports 7
Layers of the Ports' ICT System 8
Security and Safety: Two Interrelated Concepts 9
Maritime Security Organizations 9
Security of Port Services 13

3. Security of Ports' Critical Information Infrastructures **27**
Safety Management: A Restricting Approach 28
Cybersecurity Regulations and Standards 31
Security Management: A Holistic Approach 34
CIIP Methodologies 46
CYSM Risk Assessment Tool as a Best Practice 51

4. Maritime Supply Chain Risk Assessment (at Entity Level) **67**
Supply Chain Graph Models 68
Medusa: A Maritime SCS Risk Assessment Methodology 79
The Medusa SCS Risk Assessment System 87
Validation Scenarios 92

5. Maritime Supply Chain Risk Assessment (at Asset Level) 103
Standards and Methods 103
MITIGATE Risk Assessment SCS Methodology at Asset Level 106

6. Conclusions and the Way Forward 125

Bibliography 135
Appendix A: CYSM Questionnaire for Ports' Security Awareness 145
Appendix B: Threat Analysis: An Example 157
Appendix C: Supply Chain Controls and Vulnerabilities 175
Index 185

LIST OF FIGURES

Figure 1.1	Maritime logistics and supply chains.	4
Figure 2.1	Maritime environment.	8
Figure 2.2	Security (cybersecurity) and safety.	10
Figure 2.3	Vehicles transport service.	17
Figure 2.4	Vessel arrival and vehicles uploading process.	19
Figure 2.5	Execution steps of the "vehicles robbery" attack during vessels unloading.	24
Figure 2.6	Left image: mapping SCADA vulnerabilities with BPMN events. Right image: vulnerabilities of SCADA asset.	25
Figure 3.1	Collaborative cyber/physical security management (CYSM) system.	55
Figure 3.2	CYSM Administration module.	56
Figure 3.3	CYSM Management module.	58
Figure 3.4	CYSM Risk Assessment module.	58
Figure 3.5	CYSM Risk Assessment Results module.	59
Figure 3.6	CYSM Security Policy Reporting module.	60
Figure 3.7	CYSM Risk Assessment Toolkit architecture.	61
Figure 4.1	An example of an SC-directed graph.	69
Figure 4.2	The components and layers of the Medusa system.	87
Figure 4.3	Using the control implementation level, as provided in the Security Declaration Statement, to assess the vulnerability level for each threat in the Medusa system.	89
Figure 4.4	Assessment of each threat scenario that is relevant to the vehicle transport SCS in the Medusa system.	90
Figure 4.5	The risk assessment phase of the Vehicle Transport SCS in the Medusa system.	90
Figure 4.6	The cascading dependency risk assessment phase by the Medusa system.	91
Figure 4.7	An SCG based on the dependencies of the Purchase and Shipment SC.	92
Figure 5.1	MITIGATE high-level architecture.	122
Figure 5.2	MITIGATE dashboard overview.	123

LIST OF TABLES

Table 2.1	Vehicles transport supply chain service	18
Table 2.2	Threat analysis of the SCS "Vehicles' Transport"	21
Table 3.1	Assessment of Security Management Methods and criteria	42
Table 3.2	CIIP methodologies short description	47
Table 3.3	Assessment of suitable CIIP methods	50
Table 4.1	Part of a security declaration statement	70
Table 4.2	Security vulnerabilities and the corresponding security controls	71
Table 4.3	Assigning threat scenarios to threat categories	73
Table 4.4	Assigning security vulnerabilities and security controls to threat categories and their related threat scenarios	75
Table 4.5	Threat scale	76
Table 4.6	A likelihood scale	77
Table 4.7	A consequence scale	78
Table 4.8	A risk scale	79
Table 4.9	Product of likelihood values calculation	85
Table 4.10	An example of input values for the calculation of cascading risk	85
Table 4.11	Validation scenario 1: implementation of different security controls and related risk levels	94
Table 4.12	Validation scenario 2: variation of the expected consequences and related risk levels	96
Table 4.13	Validation scenario 3: variation of the probability of occurrence of the threat scenarios and related risk levels	98
Table 4.14	Validation scenario 4: cascading risks	100
Table 4.15	Dependency chains with the port authority as the destination for the threat scenario $TS_{1.1}$	101
Table 5.1	Mapping between SCRA main blocks and substeps	109
Table 5.2	Mapping of the CVSS metrics on the MITIGATE vulnerability level	114
Table 5.3	Mapping of the attacker's capability and the IVL onto the likelihood of exploitation	114
Table 5.4	Description of the probability scale in MITIGATE	115
Table 5.5	Mapping of the CVSS metrics on the MITIGATE impact level	118
Table 5.6	Mapping of the impact level and the ICVL onto the individual chain impact level	118
Table C.1	Medusa's security declaration statement	176
Table C.2	Security vulnerabilities and related security controls related with supply chain security	180

ACKNOWLEDGMENTS

The author is grateful to the European Commission (Horizon 2020 programme) for funding the maritime cybersecurity projects CYSM (CIPS 2012), MEDUSA (CIPS 2014), and MITIGATE (Horizon 2020); this book is based upon the main findings of these projects of which the author served as project/technical manager. The author also thanks all partners involved in these projects, namely:

- Port Institute for Studies and Co-Operation in the Valencian Region, FEPORTS
- University of Piraeus, Research Center
- SingularLogic
- Port Authority of Pireaus
- Università degli Studi di Genova (DITEN)
- Fundación Valencia Port
- Europhar
- Austrian Institute of Technology
- University of Cyprus
- Fraunhofer CML
- Maggioli Group
- University of Brighton

The author is also thankful to the European Union Agency for Network and Information Security (ENISA) that allowed her to contribute in the first study on maritime cybersecurity issues in 2011 entitled "Cyber Security Aspects in the Maritime Sector." Finally, the author would like to express her acknowledgment to the University of Piraeus, Research Center.

(*To my beloved mother and daughter for all their support in my life*)

GENERAL SECURITY GLOSSARY

Term	Explanation
Asset	Anything valuable for an organization (ICT, physical system, users, processes, things).
Attacker	A person or independently executing program that intends to compromise the security (confidentiality, integrity, availability, authenticity) or privacy of an asset.
Attacker profiles	Attackers will be identified by their relationship with the organization (insider, outsider), their skills (ICT skilled, premature), and their goal; each attacker will be characterized by a triple of attributes, e.g., (insider, premature, exploit specific asset), (outsider, premature, no target, general damage).
Confirmed vulnerabilities	Vulnerabilities that have not been treated, i.e., security controls have not been implemented for preventing their exploitation.
Critical information infrastructure (CII)	A physical/virtual information system that can control, process, exchange, or store electronic information that is vital to the functioning of critical infrastructure.
Impact level	Measures the effect that can be expected because of the successful exploitation of a vulnerability that resides in an asset.
International ship and port facility security code (ISPS)	Set of measures to enhance the security of ships, ports infrastructures, and facilities.
Intrusion detection system	A piece of software or device that monitors network and/or system activities for malicious activities and/or policy violations to notify the respective stakeholders.
Malware	Abbreviation for malicious software. A piece of software that is used by a malicious adversary to infiltrate, damage, and cause unauthenticated and unwanted actions on an information system of a victim.
Potential/unknown (zero-day) vulnerability	An undisclosed vulnerability that could be exploited by a threat.
Risk assessment	The procedure of evaluating (cyber, physical) the risk levels that an organization is exposed to. Depending on the method, the assessment can be qualitative or quantitative.

Continued

—*Continued*

Term	Explanation
Risk level	The quantification of the risk that an organization is exposed to, usually in the form of groups (low, medium, high).
Security	The assurance of confidentiality, integrity, authenticity, and availability.
Security authority	Person or department responsible for the thorough observation and execution of security issues and measures.
Security declaration statement	A documented commitment that specifies the security measures/controls undertaken by an organization to protect its assets.
Threats	The means by which a potential security incident (deliberate or accidental) might occur, affecting an asset(s).
Vulnerability	A weakness or a flaw in an asset, raised either from implementation, design, or other processes, that can be exploited or triggered by a threat. Vulnerabilities could be induced through poor configuration, lack of security patching, etc.
Vulnerability assessment	The procedure for estimating the vulnerability levels of all vulnerabilities of all assets under assessment.
Vulnerability level	Measures the probability that an attacker can successfully reach and exploit a specific (confirmed or zero-day) vulnerability in each cyber asset.

MARITIME GLOSSARY

Term	Explanation
Authorized economic operator (AEO)	Party complies with supply chain security standards (e.g., from World Customs Organization, WCO).
Barge operator	Company that provides barge capacity and barge transport.
Business partner	Ports/port authorities, suppliers, contractors, suppliers, service contractors involved in the provision of a supply chain service (SCS) or in any process/subprocess of the SCS.
Carrier	Freight transporting company.
Combined transport	Intermodal transport where the major part of the European journey is by rail, inland waterways, or sea and any initial and/or final legs carried out by road.
Confirmed vulnerability	Confirmed vulnerabilities are these vulnerabilities that have not been treated, i.e., security controls have not been implemented for preventing their exploitation.
Consignee	Ultimate receiver of the shipper consignment.
Consigner	Sender/owner of the transported goods, puts goods in the care of a forwarding agent or transport operator.
Customs	Authority or agency in a country responsible for collecting tariffs and for controlling the flow of goods.
Depot	Warehouse designed for the loading and unloading of goods directly from trucks, railways, or seaports.
Distributor	An entity that buys noncompeting products or product lines and resells them to retailers or directly to the end users or customers.
Dry port	Inland terminal that is directly linked to a maritime port.
Exporter	A business person or firm who transports goods abroad for sale.
Feeder service	Short sea shipping service that connects at least two ports to consolidate or redistribute freight from a deep sea service.
Ferry operator	Company operating and/or owning ferry vessels.
Feeder vessel operator	Company operating and/or owning vessels that is specialized in feeder operations.

Continued

—*Continued*

Term	Explanation
First-order logic	A formal system for symbolized reasoning in which a statement is decomposed into terms, and atomic statements to validate the truthfulness of a statement.
Forwarder	Company that is specialized in carrying out the forwarding function on behalf of a shipper. The company can either carry out that forwarding function or can additionally offer further logistical services and/or owns rolling road stock.
Harbor	Area of water providing shelter for ships, used to build ports.
Harbor master	An official, responsible for enforcing the regulations of a particular harbor or port, to ensure the safety of navigation, the security of the harbor, and the correct operation of the port facilities.
Importer	Firm or person whose business involves importing goods from outside (especially from a foreign country).
Inland carrier	A transportation line that hauls cargo inland: truck, rail, barge, inland waterways, or domestic airline flights.
Intermodal operator	Company that offers the organization of either a door-to-door or a terminal-to terminal transportation chain by using the means of intermodal transport. Intermodal operators are those actors that combine and refine the crude rail haulage capacity of individual railways, adding to it specific support services and making it into a saleable product. Their status may be a railway, or railway-owned or largely private as the UIRR companies.
Intermodal transport	Movement of goods in one loading unit or road vehicle, which uses two or more modes of transport.
Lift-on/lift-off	Loading and unloading of intermodal transport units using lifting equipment.
Loading/unloading port	A place on a waterway with facilities for loading/unloading ships.
Logistics	The process of designing and managing the supply chain.
Logistic service provider	A company that offers logistic services like warehousing, storage, stuffing, and stripping, etc., but in any way services other than just transport and forwarding.
Maritime supply chain	Handling of goods toward and from ports as well as transport by ship including handling and packaging operations.

—*Continued*

Term	Explanation
Maritime stakeholders	All organizations, companies, and persons involved in enabling and conducting maritime transport as well as in shaping of maritime policies and directives concerning the maritime supply chain.
Multimodal transport	Carriage of goods by two or more modes of transport.
Multimodal transport operator	Person who concludes a multimodal transport contract and assumes the whole responsibility, performs as carrier or transport operator.
Operator	Represented in the transport chain as forwarder, intermodal operator, agent, or terminal operator having each of their functions either to plan and/or to control each transport stage and/or load unit handling in the terminals.
Port authority	Governmental or quasigovernmental body that operates ports and other transportation infrastructure.
Policy actor	Representative of a group of interests within a country community, a country, or a region, influencing decisions of transport actors concerning the choice of the means of transport by policy frameworks such as fiscal and order policy measures.
Port community system (PCS)	Electronic platform that includes different software modules for logistic operations in the direct port environment as well as in connected subsystems enabling intelligent and secure exchange of information between public and private stakeholders.
Road haulier	Company operating and/or owning trucks and carrying out road transport functions concentrating on the physical movements of goods.
Roll-on/roll-off	Loading and unloading of a road vehicle, wagon, or intermodal transport unit on or off a ship on its own wheels or wheels attached to it for that purpose.
Shipper	Owner of the cargo when it is dispatched, can be either a consignee or a consignor.
Ship agent	A person or company that carries out the functions of an agent irrespective of whether they are in business as a ship agent or they perform such functions as an adjunct to, or in conjunction with, other activities such as ship owning or operating, providing cargo handling, or similar. Person or firm that transacts all business in a port on behalf of ship owners or charterers.

Continued

—*Continued*

Term	Explanation
Shipping line/coastal/ oversea shipping line	A company operating and/or owning vessels for sea transport.
Short sea shipping	Movement of cargo by sea within Europe or between Europe and North Africa.
Stevedoring company	Company responsible for the storage of goods at terminals and storing on vessels.
Supply chain	Supply chain is the linked set of resources and processes that begins with the sourcing of raw material and extends through the manufacturing, processing, handling and delivery of goods, products and/or services to the consumer through different transport means.
Supply chain service	Service provided by a supply chain, a linked set of resources and processes.
Terminal	A facility where cargo is trans-shipped between different transport modes, for onward transportation.
Terminal operator	Company organizing and conducting all handling and administrative processes on a terminal.

EXECUTIVE SUMMARY

The maritime ecosystem is complex and involves many entities that interact with each other. Examples of these entities are ports, ships, port authorities, maritime and insurance companies, customs, the ship industry, banks, ministries, other commercial providers, and other infrastructures (e.g., railroads, airports). All these interactions are supported by complex and heterogeneous information and communication technology (ICT) systems.

Commercial ports are among the transportation critical infrastructures since they are large-scale infrastructures of which the degradation, interruption, or impairment of their physical or cyber (ICT) systems has serious consequences on national security, health, safety, economy, and welfare of citizens and nations, and they are characterized by multiplicity of interdependencies with other entities in the maritime ecosystem.

The normal functionality of the commercial ports depends largely on the proper operation of both their physical and cyber systems. The large amount of critical and sensitive data, the information and services that are managed daily, the large number of entities called to be served, and the interdependencies with the other infrastructures require effective security management.

This book explores the existing picture in the security of the commercial ports' critical information infrastructures (CIIs) and their supply chains and goes a step forward by identifying gaps in the security management of the ports' CIIs, drawing proposals, research directions, and recommendations.

CHAPTER 1

Introduction

Commercial seaports are considered critical facilities for a country's economy since any disruption of their services could cause damages in the economy, to the safety of people, and a lack of basic services. Given their position in coastal areas and the great variety of substances handled there, ports (especially those of a certain size) are now considerably complex systems from an environmental and safety point of view. They tend to be associated with water and air pollution, soil contamination, problems related to dust and noise, the generation of waste, dredging operations, movements of ships, trucks, and trains, warehouse storage of hazardous substances, etc. In addition, ports are usually close to urban centers. In fact, for historical and practical reasons, many ports are closely linked, physically and socially, to a city, and some of them are physically very close or belong to a certain city. These circumstances mean that the impact on the environment and people from certain accidents can be very serious. Given the properties of some substances that pass through the ports (chemical products, hydrocarbons, fertilizers) and the operations that are carried out on them (loading and unloading, storage, transport), the possibility of there being an accident is hardly negligible. In fact, there are periodic fires, explosions, and toxic releases, with possible consequences of an impact on the environment, alarm among the population, financial losses, etc. In addition to these direct consequences, another quite important aspect of these situations is the negative image they give to the port, with the possible creation of a feeling of rejection among the population. On the other hand, it must be considered that, in the latter circumstances, the possibility of moving the port facilities further away from the population is, in most cases, practically impossible. To tackle the hazards, prevent damages, and manage consequences, ports have implemented safety management systems tailored to the specific nature of their activities. Furthermore, national and international rules have been issued by governments and international organizations for facing risks on ships, ports, and maritime operations.

First, management systems and legal frameworks were addressed to ships and navigation. The Safety of Life at Sea (SOLAS) Convention in its successive forms is generally regarded as the most important of all international treaties concerning the safety of merchant ships. The first version was adopted

Port Cybersecurity
ISBN 978-0-12-811818-4
http://dx.doi.org/10.1016/B978-0-12-811818-4.00001-0

in 1914, in response to the *Titanic* disaster. The tragic accident of *Titanic* (1912) imposed a new concept of "safety" in the maritime sector, and numerous legislations and directives have been published since then concentrating on the physical protection of the ships, crew, passengers, cargo, and sea.

The second version of SOLAS was in 1929, the third in 1948, and the fourth in 1960. The 1974 version includes the tacit acceptance procedure, which provides that an amendment shall enter force on a specified date unless, before that date, objections to the amendment are received from an agreed number of parties. As a result, the 1974 Convention has been updated and amended on numerous occasions. The convention in force today is sometimes referred to as SOLAS, 1974, as amended. The main objective of the SOLAS Convention is to specify minimum standards for the construction, equipment, and operation of ships, compatible with their safety. Flag states are responsible for ensuring that ships under their flag comply with its requirements, and several certificates are prescribed in the convention as proof that this has been done. Control provisions also allow contracting governments to inspect ships of other contracting states if there are clear grounds for believing that the ship and its equipment do not substantially comply with the requirements of the convention; this procedure is known as port state control. The current SOLAS Convention includes articles that set out general obligations, amendment procedure, etc., followed by an annex divided into 12 chapters. Even though there are some chapters dedicated to carriage of cargoes, dangerous goods, and management of safe operations, they all are directly related to ships and not to port facilities or the interface ship/port. Port facilities are considered working places, industrial areas, storage, production, and handling areas that fall under general rules and legislation issued by the countries and international organizations.

The terrorist events in New York and Washington (2001), Madrid (2004), and London (2005) imposed the concept of "security" with additional directives and legislation (e.g., the International Ship and Port Facility Security Code). Although these new efforts concentrated on the organizational and auditing aspects of the physical security of the ships, maritime companies, and commercial ports, they have not considered the information and communications technology (ICT) and cybersecurity threats and risks. The shipwreck in Italy (2012) was a natural consequence of this ignorance since it was found that the communication between ports and ships was not reliable; the navigation and coastline systems did not function properly, and a disaster recovery plan was not implemented. In other words, many aspects of cybersecurity were not covered.

In December 2011, European Union Agency for Network and Information Security (ENISA) [33] published a report analyzing the cyber-security challenges in the maritime sector. This report presented the results from the ENISA workshop [35] on this topic concentrating on the ICT port security and specifically in the security management of ports information and communication technology systems. The surprising observation as stated in the report was that "*the awareness on cyber security needs and challenges in the maritime sector is currently low to non-existent.*"

In the modern era, ports' critical infrastructures (CIs) and maritime logistics and supply chains are the blood veins of global trade and economy where cross-border CIs (e.g., ports, maritime authority's airports, railways, energy providers, banks, maritime/logistic/transport companies) collaborate in offering critical complex services (e.g., container management, vehicle transport, liquefied natural gas [LNG] storage and transport, cruising). The CIs that operate within these services have physical and cyber multiinterdependencies, interacting with all sectors of the economy; therefore, their malfunctioning or disruption will have cascading effects on several other infrastructures or cross-border services that depend on them.

Most physical processes within a maritime service (e.g., vehicles and cargo loading/unloading, LNG distribution and storage) are executed with autonomous or semiautonomous mechanical physical systems and machineries (e.g., ships, trucks, cranes, electronic gates/fences) under the control of sophisticated logistic software systems (e.g., Industrial Cyber-Physical Systems, supervisory control and data acquisition (SCADA), surveillance systems). In addition, cargo ships relate to the ports and the other operators via a plethora of communication and data links (e.g., satellite communications or conventional radio communications), and their navigation is today widely reliant on electronic solutions (e.g., satellite navigation with GPS, Galileo, or Electronic Chart Display Information Systems [ECDIS]). Thus, maritime supply chains can be viewed as complex physical–cyber systems composed of heterogeneous, interconnected physical and cyber assets, owned by different national, European Union (EU), and international CIs, ensuring seamless and swift product/data exchange from the producer down to the end consumer during the provision of these services.

The maritime, highly valuable, interconnected physical and cyber assets (Fig. 1.1) have become targets for attacks, attracting the attention of terrorism (e.g., Russian Business Network), cyber hacktivism organizations, militias (e.g., Anonymous, LulzSec), and agencies (e.g., Stuxnet, Flame, Conficker, DuQu, APT1). In particular in the last 3 years, they have garnered

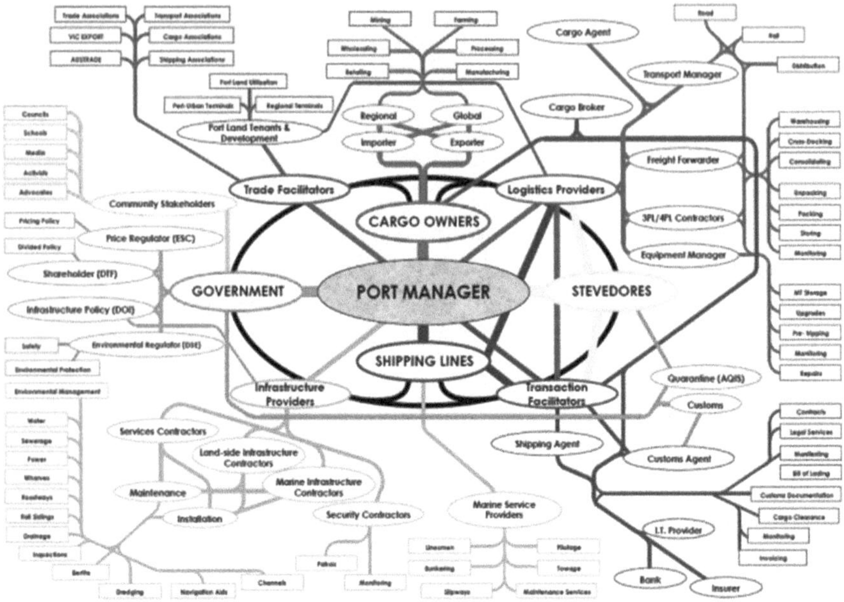

Figure 1.1 Maritime logistics and supply chains.

front-page attention as victims of cyberattacks: (1) a Chinese manufacturer implanted malware in inventory scanners to steal supply chain intelligence; (2) hackers shut down a floating oil rig by tilting it; (3) Somalian pirates used low-cost GPS jammers to change a vessel's course by interfering with its navigation systems (GPS, ECDIS, and AIS), causing a track line–following autopilot to inaccurately interpret the ship's position and alter its course; (4) hackers gained access to a shipping company's databases and vessel tracking systems to identify vessels with valuable cargo; thus, many ships that transit the Gulf region are turning their AIS navigation tracking system off, so pirates cannot identify, locate, and track them; (5) in port of Antwerp, between 2011 and 2013, cyberattacks were used to highjack, divert, and steal cargo.

Attacks on the maritime services cause not only disruption of the services but tremendous damage to the maritime operations, national and EU safety, economies, societies, and environment. For example, attacks in the Industrial Control Systems (ICSs) (e.g., supervisory control, SCADA, distributed control systems, and programmable logic controllers) hosted in ports or maritime transport companies may cause disruption or damage of critical mechanical devices (e.g., container cranes, safety and mechanical systems that operate locks and dams), and even worse, they may cause loss of life, stealing of cargo, or destruction of ships. The effects (in terms of thermal radiation, overpressure blast wave, and flying shrapnel) of the explosion of an LNG tanker or in the port's LNG storage facilities or terminals due to a hacked SCADA system could lead to a lack of energy stock, which could be critical during cold waves, affecting the environment (degradation, fragmentation, or loss of ecosystems), the economy, and more importantly, the wellness and health integrity of the citizens. An attack on a container terminal management system could disrupt intermodal container services involving maritime, rail, and truck transportation. Cyberattacks (e.g., inserting a malware) in the port's SCADA systems may cause fuel spills affecting water quality; attacks in the Port Community Systems (PCS) may turn LNG tankers into floating bombs; physical attacks (e.g., bombing) in a dry bulk storage area of coal products may create and carry dust by wind to tourist terminals or nearby residences. Modern surveillance systems, monitoring units, sensors, RFIDs, marine traffic systems, PCS, and automated border control systems become more and more susceptible to current threats; they are prone to snooping, interruption, vandalism, and interception that may cause the disruption of various critical maritime operations and services and, even worse, to illegal movement of humans and goods, human and drug trafficking, and terrorism. It should be noted that according to 2016 estimates by the RAND Corporation and the USA Congressional Research Service, an attack on a port's CI could cause tens of thousands of deaths and cripple global trade, with losses ranging from $45 billion to more than $1 trillion.

There is an urgent, pressing challenge for the ports and maritime security officers and operators to protect their interconnected CIs' physical and cyber systems in the new digital maritime era.

The International Port Community Systems Association in 2015 recommended that all maritime, logistics, supply chain actors, and PCS operators with the support of their trade associations and international and regional bodies address the current threats to the supply chain, bring together key stakeholders in the supply chain, and create an information sharing and analysis center to discuss cybersecurity threats, risks, and experiences.

In 2016, International Maritime Organisation developed guidelines for maritime cybersecurity management to provide high-level recommendations to safeguard shipping from current and emerging cyber-related threats and vulnerabilities.

In 2016 the House of Representatives in the United States introduced Bill H.R. 3878, "Strengthening Cybersecurity Information Sharing and

Coordination in Our Ports," and finally, the EU Directive NIS 2016 enforces all CIs to report to an appropriate Computer Security Incident Response Team any incident having substantial impact on the provision of their services.

This book aims to contribute toward the answers of the following questions:

1. What are the threat and vulnerabilities in the ports and supply chains?
2. What are the relevant efforts of the maritime community related to security? What are the gaps and open issues?
3. Which security management standards and methodologies can be applied in ports' CIs and their supply chains, and which modifications are required?
4. How can European Critical Infrastructure standards and directives serve as guidelines to the enhancements of commercial ports and their supply chain security?
5. What are the threats in a maritime supply chain? How can vulnerabilities propagate in a maritime supply chain?
6. Security management tools can effectively be used to manage, monitor, and audit the ports and their supply chain infrastructures. Can we find targeted security management tools?

In this book, commercial ports are treated as critical information infrastructures (CII), thus hosting critical physical/cyber systems, offering critical services, and imposing additional security requirements (e.g., confidentially integrity, authentication, and availability of data and users).

The targeted audience of this book is the maritime scientific community, i.e., maritime security providers, ICT integrators, port authorities, supply chain business partners (maritime/transport/logistics companies), and maritime standardization bodies, organizations, and associations.

CHAPTER 2

Ports' Critical Infrastructures

As defined by the European Council (2008) a **critical infrastructure** (CI) is an "asset, system or part which is essential for the maintenance of vital societal functions, health, safety, security, economic or social well-being of people, and the disruption or destruction of which would have a significant impact because of the failure to maintain those functions."

In today's societies, the advanced development of information systems and the diffusion and fast evolution of broadband communication infrastructures has caused the wide adoption of information and communication technology (ICT) by all CIs. The degradation, disruption, or destruction of these infrastructures would have a serious impact on the daily life of the citizens. Subsequently, information processes supported by ICT that form critical infrastructures for themselves or that are critical for the operation of other CIs are defined as *Critical Information Infrastructures* (CIIs).

Commercial ports are transport CIIs in the digital era since they host critical ICT systems, the disruption or destruction of which has a significant impact on economy, commerce, and national safety [5]. The normal functionality of the commercial ports depends largely on the proper operation of their ICT systems. The large amount of critical and sensitive data, the information and services that are managed daily, the large number of entities called to be served, and the interdependencies with the other infrastructures require effective security management.

Commercial ports implement critical e-port/supply chain services and complex business processes, thus appropriate security management and protection approaches are required, preserving the secure performance of the ports in case of failures, attacks, accidents, or malicious acts and at minimizing the recovery time and damage. Major attention should be given to the cybersecurity of existing, developing, and new generation critical e-port and supply chain services (SCSs).

MARITIME ENVIRONMENT: THE ROLE OF COMMERCIAL PORTS

The maritime environment is a complex one, as shown in the next figure (Fig. 2.1), involving many entities—ports, ships (with passengers, crew, cargo), port authorities, maritime and insurance companies, customs, ship

Port Cybersecurity
ISBN 978-0-12-811818-4
http://dx.doi.org/10.1016/B978-0-12-811818-4.00002-2

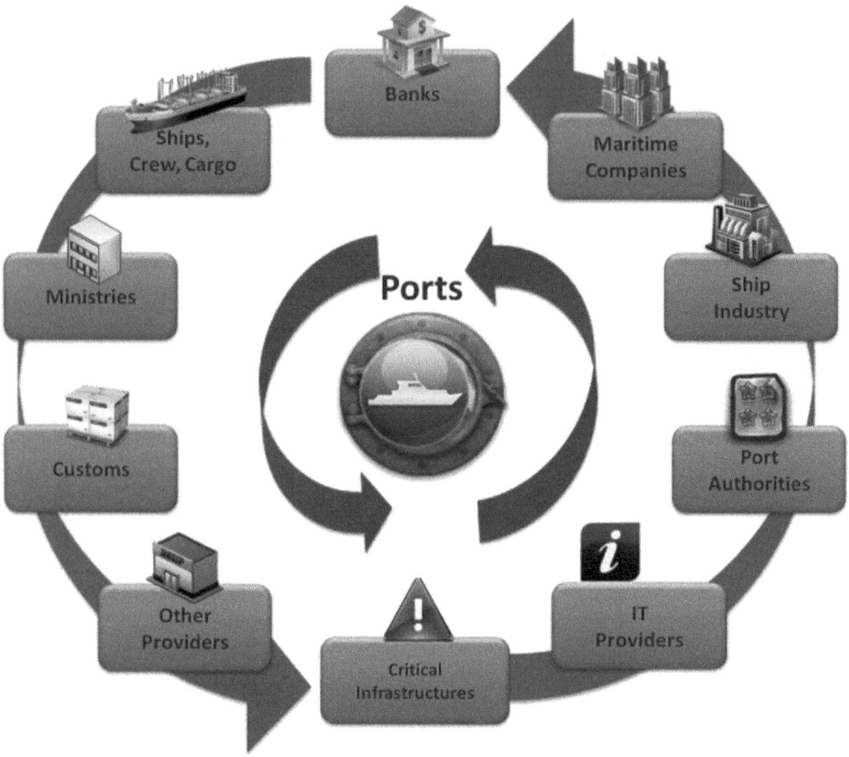

Figure 2.1 Maritime environment.

industry, banks, ministries, other commercial providers, and other CIs (e.g., railroads, airports)—that interact with each other and are supported by complex and heterogeneous physical and cyber (ICT) systems.

A central role in the maritime environment is undertaken by the commercial ports since it is the only entity that interacts directly with all maritime participants offering services with different degrees of criticality.

LAYERS OF THE PORTS' ICT SYSTEM

Many of the services offered by and processes supporting the operation of a commercial port have been automated. The **Ports' ICT systems (cyber-system),** as all ICT systems, consist of the following six consecutive layers:

1. **physical** (e.g., buildings, platforms, gates, marinas, data centers, platforms, storage rooms);

2. **network** (e.g., Internet/satellite/Wi-Fi/ad-hoc networks, telecom equipment, satellites, servers, relay stations, tributary stations);
3. **systems** and **software** (e.g., port community systems, transmission systems, data identification, maritime navigation, enterprise resource planning, ticketing, geographic information systems, port resilience systems, software);
4. **information** and **electronic data** (e.g., marine and coastal data, trade data);
5. **services** (e.g., invoicing, navigation, container management, vehicle transport luggage/cargo/vessel management, logistics, e-health, energy services);
6. **users/processes**: This layer consists of all users that interact with the preceding layers, i.e., internal users (e.g., administrators, personnel); external users (e.g., port authorities, maritime companies, customs, insurance companies, ICT, and commercial provides); things (e.g., ships, crew cargo, luggage, vehicles). Processes and procedures (e.g., loading/uploading) followed for the operation of the ports are also included in this layer.

The ports' ICT systems can be hosted in one building or can be distributed in various locations of the ports' CIs. Various threats (see an exhaustive list in Appendix B) can target all assets at all layers of an ICT system.

SECURITY AND SAFETY: TWO INTERRELATED CONCEPTS

It is very common in the maritime security literature that the concepts of security and safety are misinterpreted or mixed. For this reason, we provide some clarifications for the meaning and the interrelationship between these two concepts.

We say that an ICT system is **secure (cybersecure)** if all assets in the six layers of the ICT system satisfy all four dimensions of security, i.e., confidentiality, integrity, authenticity/access control, and availability. By **safety** (or physical security), we mean the satisfaction of the two components of security, namely, access control and availability of the assets in the first and sixth layers. Therefore, safety is a subset of ICT security, as illustrated in Fig. 2.2.

In the maritime literature, there is a misunderstanding about these two concepts, and they coincide in most cases.

MARITIME SECURITY ORGANIZATIONS

There is an increasing number of organizations that deal with the safety of the ports' CIs, which recognize the importance of their security (cybersecurity) as well.

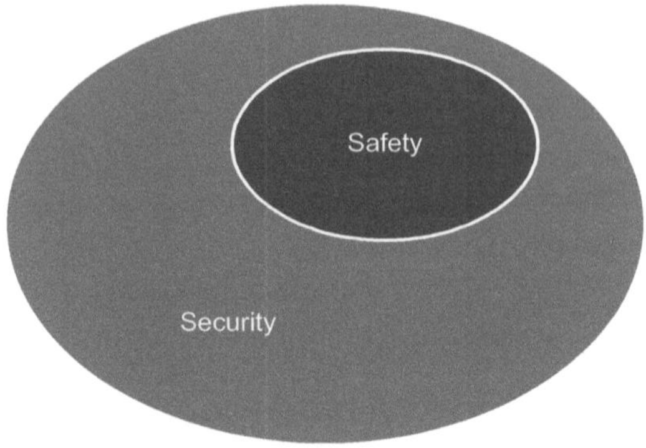

Figure 2.2 Security (cybersecurity) and safety.

International Maritime Organization (IMO) [149]: Since its establishment in 1959, IMO has adopted about 50 international conventions and protocols and has released over 800 safety codes, recommendations, and guidelines. Most of these conventions and protocols fall into the following categories:

- *International Convention for the Safety of Life at Sea (SOLAS)* [57]: Its objective is to define standards for the ships' construction, equipment, and operation, compatible with their safety.
- *International Convention for the Prevention of Pollution from Ships (MARPOL)* [56]: This is the international convention that covers aspects relating to the prevention of pollution of the marine environment by ships from operational or accidental causes.

The 2002 IMO directive ISPS (International Ships and Port Facilities Security Code) is the first one that addresses security issues, but it is restricted in the organizational schemas of security and the access control/authentication measures.

In 2016, IMO issued the cybersecurity guidelines to provide high-level recommendations to protect shipping from current and emerging cyber threats and vulnerabilities; emphasis is provided in the need to perform cybersecurity risk management in the shipping systems to estimate the impact of the cyber threats and evaluate the cyber risks.

NATO's Alliance Maritime Strategy specifies various activities that will contribute toward secure shipping by organizing various activities in collective defense, crisis management, cooperative security, and maritime security. In 2016, NATO and the European Union (EU) signed an agreement to collaborate on cybersecurity defense [37]. The EU and NATO signed in

2016 a Technical Arrangement between the NATO Computer Incident Response Capability (NCIRC) and the Computer Emergency Response Team–European Union (CERT-EU) facilitating the information sharing between NCIRC and CERT-EU to improve cybersecurity management in both organizations, improving the cyber defense between the EU and NATO. The NATO Maritime Interdiction Operational Training Center has organized a series of workshops and training sessions on maritime cyber security issues, enabling the collaboration with cybersecurity experts and professionals.

The **European Maritime Safety Agency** (EMSA) was created in 2004, under Regulation (EC) No 1406/2002 establishing a European Maritime Safety Agency, for ensuring a high, uniform, and effective level of maritime safety and prevention of pollution by ships within the community. The agency provides the member states and the commission with the technical and scientific assistance needed and with a high level of expertise, to help them to apply community legislation properly in the field of maritime safety and prevention of pollution by ships, to monitor its implementation, and to evaluate the effectiveness of the measures in place. EMSA basically does the following:

- ensures that all EU member states implement correctly European regulations in the field of maritime safety;
- supports the European Commission in its legislative task and international forums, such as the IMO;
- builds and operates common tools used for maritime safety tasks by member states (e.g., SafeSeaNet for vessel traffic monitoring, CleanSeaNet satellite images for identifying pollution at sea, EMCIP for centralizing data on marine accidents, training tools, etc.);
- creates stronger links between national maritime administrations, by hosting meetings, conducting training, and sharing best practice;
- stimulates cooperation and training linked to maritime safety issues.

The **European Sea Ports Organisation** (ESPO) was founded in 1993. It represents the port authorities, port associations, and port administrations of the seaports of the member states of the EU and Norway. ESPO has also observer members in several neighboring countries to the EU. ESPO ensures that seaports have a clear voice in the EU. The organization promotes the common interests of its members throughout Europe and is also engaged in dialogue with European stakeholders in the port and maritime sector.

ESPO's mission is to influence public policy in the EU to achieve a safe, efficient, and environmentally sustainable European port sector, operating as a key element of a transport industry where free and undistorted market conditions prevail, as far as practicable. ESPO's key objectives are these:

- to ensure the economic importance of European ports;
- to promote free and fair competition in the port sector;
- to ensure that European ports play their full part in delivering economic efficiency;
- to promote the highest possible safety standards in European ports;
- to encourage ports to be proactive in protecting the environment.

With respect to safety, EMSA is specialized in safety guidelines and training related to ship operations at ports, e.g., liquid natural gas bunkering, waste reception, loading and unloading of general goods, containers, and bulks, and container movements. In the 2016 annual ESPO conference the need for security plans to mitigate cyber risks was among the main topics.

Innovation and Networks Executive Agency (INEA) is the successor of the TEN-T EA [154] aims that aimed to support the European Commission, by ensuring the successful implementation of the Trans-European Transport Network (TEN-T) via various projects dealing with the following:

- *Inland waterways* aimed to strengthen the bond between inland regions and coastal areas (e.g., ports) connected via links such as rivers and canals.
- *River Information Services* emphasized the optimization of the traffic management infrastructure on the inland waterway network via the establishment of an interoperable, intelligent traffic and transport system.
- *Seaports* were aimed to the development of sea transport via the design, implementation, and operation of a set of services for passenger and goods transport, including ferry services and short- and long-distance shipping services, coastal shipping, linking EU Member States together and with third countries.
- *The trans-European network of Motorways of the Sea* focused on the development of the appropriate port facilities, infrastructure, and services recreating the road and rail network on the water.

INEA is involved in the Horizon 2020 Programme with the aim to develop projects that will improve the development of smart, green, and integrated transport and encourage secure, clean, and efficient energy.

The **Maritime Security Council** [150] established in 1988 addresses several security concerns of the United States and international maritime community. It cooperates with a set of US government agencies concerned with maritime security and counterterrorism and acts as the maritime advisor of the US government and the international police agency, INTERPOL.

Its main objective is to support the international merchant marine community on issues relating to criminal and terrorist threats through the following channels:

- representing maritime interests before government bodies;
- acting as liaison between industry and government;
- disseminating timely information; encouraging and assisting in the development of industry-specific technologies;
- convening educational and informational conferences for its membership and government partners.

It works to prevent illegal drug trafficking, stowaway, theft, privacy, terrorism, and hijacking. However, it should be noted that the agency does not deal directly with ICT or cybercrime issues.

The **European Network and Information Security Agency** (ENISA) is a center of expertise for the EU, its member states, the private sector, and Europe's citizens. As an EU agency, ENISA's role is to work with these groups to develop advice and recommendations on good practice in information security. The agency assists member states in implementing relevant EU legislation, and it works to improve the resilience of Europe's critical information infrastructure and networks. In carrying out its work program, ENISA seeks to enhance existing expertise in member states by supporting the development of cross-border communities committed to improving network and information security throughout the EU. In 2011, ENISA published the first maritime cybersecurity report "Cyber Security Aspects in the Maritime Sector" acknowledging the low awareness level of cybersecurity in the maritime sector, providing recommendations and guidelines for cybersecurity awareness.

SECURITY OF PORT SERVICES

There are various port services that can be categorized according to their main goal and functions into the following five categories:

- **vessel management services**: providing e-information about vessel status for all agents; e-administrative procedures; e-communication with port/maritime authorities, police, immigration, etc., vessel authentication and monitoring services (via RFIDs, geographic information systems); navigation services;
- **cargo management services**: providing e-documentation to all agents involved; e-information about container status, cargo e-management, cargo authentication and monitoring services (via RFIDs, geographic information systems); navigation services;

- **inland logistics services**: e-management of inland transport operations (e.g., transport/accept/deliver e-orders); e-procurement, e-invoicing, e-payment; e-tracing services (e.g., e-information about shipments across entire logistics chain); e-reservation;
- **communication at port level**: multichannel communication (satellite, Internet, Wi-Fi, VPNs, ad-hoc networks) services in the port for communicating with all maritime entities (other ports, ships, crew, etc.);
- **integration with other systems**: integration services with customs systems for e-administrative various customs documents (e.g., customs declarations, import/export) and controls (e.g., taxes, penalties); integration with port police and immigration systems for e-monitoring and auditing vessels, cargo, goods, and people; integration with health organizations systems for offering e-health services (especially to the crew); integration with other transport CIIs (e.g., railways, airports) offering collaborative e-tourism services (reservation, scheduling, ticketing).

However, not all aforementioned services have the same degree of "online sophistication," i.e., their business processes do not have the same degree of dependence upon electronic means. To measure the level of online sophistication maturity of any e-service, someone can adopt the four-level framework used in the e-government sector [38], namely, information, interaction, two-way interaction, and full-handling. The *first online sophistication* level comprises those services that provide only online information (no interaction is allowed). In the *second level,* there is the possibility to download an e-document (no further interaction is allowed). Services of the *third level* (two-way interaction) concern those that offer the possibility of an electronic intake with an official electronic form to start the procedure to obtain this service. Last, in the *fourth level*, services offer the possibility to completely treat the port service electronically, including decision and delivery. No other formal procedure is necessary for the applicant via "paperwork." The last two levels require authentication of the user and the e-port service administrator.

An e/m-port service is *business critical* if its degradation, disruption, or destruction has serious impact on the effective functioning and operation of the port and/or its interacting maritime entities as well as on the economy and governments.

Going one step beyond the state of the art, let us propose a general methodology to estimate the business criticality of e-port services. These estimates are important in the evaluation of the impact that a security violation of critical e-port services would have, focusing on the way it affects not only themselves but also the interacting maritime entities. The proposed framework comprises the following steps:

- Determine the level of online sophistication offered by the service based on the adopted four-stage framework (described earlier).
- Specify the scope of the service, whether provided at a local, national, regional European, or international level.
- Determine its dependencies with other entities or services (e.g., using business process modeling or graph theory).
- Specify the recipients of the service.
- Determine the business impact level of the service for all recipients and all dependent entities.

Adopting this simple framework, one may classify its offered services in terms of business criticality.

Commercial ports host, store, and exchange information data using various business processes (electronic, paper-based, and both) to offer electronic and mobile (e/m) *port services* to the maritime participants with the main objective, the facilitation of daily activities and interactions with them, strengthening their business and commercial activities and interacting more effectively with all maritime entities.

An e/m-port service is characterized as *security critical* if it is business critical and its degradation, disruption, or destruction has serious impact on the secure operation of its port ICT systems and/or other interconnecting ICT systems, affecting the economy or the national safety. Depending of the e/m-port service online sophistication level and its business criticality, we can assign the various security levels:

Security level 0: first-level online e-port services with low business impact. Disruption of such services has neither impact on security or privacy or business, while there are no critical dependencies to other services.

Security level 1: second-level online e-port services with low to medium business impact oriented to facilitate users from the maritime environment (not only nationally but at a European or international level) with their transactions. Disruption of such services has no impact on security or privacy, while there are no critical dependencies to other services.

Security level 2: second- and third-level online e-port services with low to medium business impact oriented to support users from the maritime environment (not only nationally but at a European or international level) with their transactions. Disruption of such services has a mere impact on security or privacy, while critical dependencies to other services are possible.

Security level 3: second- to fourth-level online e-port services with medium business impact provided to maritime entities by automating traditional business processes, rendering them as the main transaction channel with governments (not only nationally but in a European/international level) for a service whose disruption would have a crucial impact on either security or privacy or with critical dependencies to other services.

Security level 4: third- to fourth-level online e-port services provided to maritime entities with medium to high business impact by automating traditional maritime processes, rendering them as the main transaction channel (not only nationally but in a European and international level) for a service whose disruption would have a crucial impact on security and/or privacy issues and/or with critical dependencies to other services and maritime entities.

Security level 5: third- to fourth-level online e-port services with high business impact provided to maritime entities, by automating traditional maritime, business, and governmental processes, rendering them as the main transaction channel (not only nationally but in a European or international level) for a particular service whose disruption would have a crucial impact on security and/or privacy issues and/or with critical dependencies to other maritime entities and especially other CIs.

Most of the port services involve various entities formulating complex supply chains (SCs). An SC, according to ISO2800, is the linked set of resources and processes that upon placement of a purchase order begins with the sourcing of raw material and extends through the manufacturing, processing, handling, and delivery of goods and related services to the purchaser. By an SCS, we mean the service provided and/or supported by an SC.

The main port services are SCSs. Examples follow.

The "Vehicles Transport Chain" SCS (Fig. 2.3) can be viewed as a massively complex system with numerous players, including ports, shippers, and transport operators that involve the shipment and receipt of various types of vehicles and equipment such as trucks, vans, truck trailers, threshing machines, etc. This service is a relatively long and complicated process that involves domestic and international transportation, warehouse management, order and inventory control, materials handling, import/export facilitation, and information technology. In this framework, the vehicles transport affects many sectors along the SC.

The "Vehicles Transport Chain" SCS breaks down into several processes (Table 2.1) that involve several physical (docking of the ship, stevedoring,

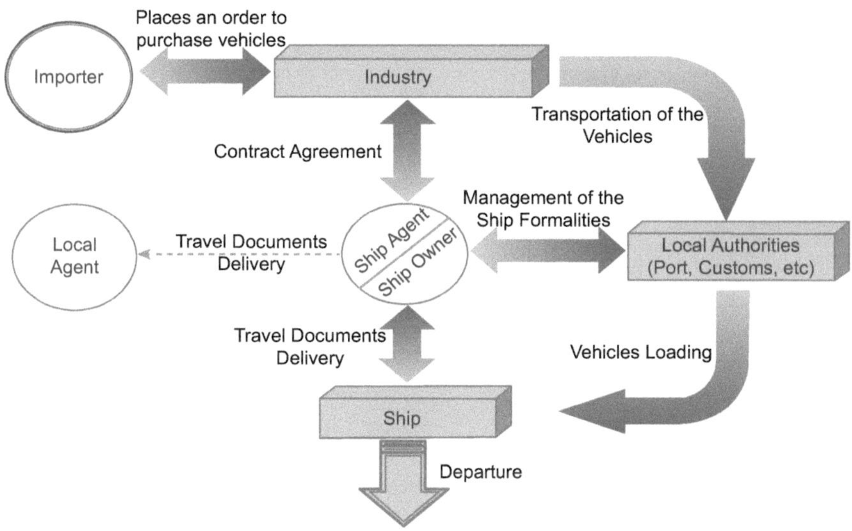

Figure 2.3 Vehicles transport service.

loading, unloading, storage, transportation, inspection, etc.) and cyber (pre-arrival notifications, customs clearance documentation management, ISPS declaration, etc.) operations, interconnections, and assets.

The analysis becomes even more complex when we concentrate on specific phases. For example, let us analyze further the "Vessel Arrival and Vehicles Uploading" phase (Fig. 2.4) of the Vehicles Transport SCS. This phase is a representative example of a complex, distributed, and interconnected process that is initiated by the ship agent that has to arrange various details about the vessel docking and unloading of the vehicles at the port terminals interacting with customs authority and public administration. The ship agent uses the e-services offered from the customs that are accessible via the Integrated Customs Information System (ICIS) to submit the ship-associated documentation requesting the customs clearance. The customs clearance is the documented permission given by the customs to import or export vehicles and proves that all customs duties have been paid and shipment procedures have been approved.

Once the ship agent obtains the customs clearance approval, he submits it to the port authority requesting services for the vessel, such as mooring, lacing, assigning risk assessment processes, weather conditions, navigational warnings, procedures for communication failure, fenders, personnel (truckers for transferring the vehicles from the ship to storage area, etc.).

Table 2.1 Vehicles transport supply chain service

Phases	Business processes	Business partners
A. Purchase phase	A.1 Vehicle order dispatch	Importer Industry
	A.2 Contract agreement on the vehicle purchase	Importer Industry
B. Shipment phase	B.1 Shipment contract preparation and negotiation process	Industry Ship owner Ship agent
	B.2 Manifest issuance	Ship owner Ship agent Customs
	B.3 Manifest registration	Ship owner Ship agent Customs Carrier agent
	B.4 Ship formalities arrangements	Local agent Customs Ship agent Port authority
	B.5 Vehicle loading	Port authority
	B.6 Shipping arrangements	Ship owner Ship agent Local agent Importer
C. Prearrival phase	C.1 Port's services requested	Port authority Ship agent
	C.2 Insurance contract preparation and negotiation process	Ship agent Ship owner
	C.3 Vehicle claims handling	Ship agent Local agent
D. Delivery phase	D.1 Vehicles unloading	Port authority Carrier agent
	D.2 Vehicle final transportation procedures	Importer Carrier agent

This process is performed via the Port Community System (PCS), which is an electronic platform that connects the multiple systems operated by a variety of organizations involved in the port's SC. This system acts as a centralized system that supports a variety of e-port services aimed at facilitating the secure and efficient electronic exchange and management of information between the involved public and private stakeholders and promoting the automatization and the smooth operation of the port and logistics

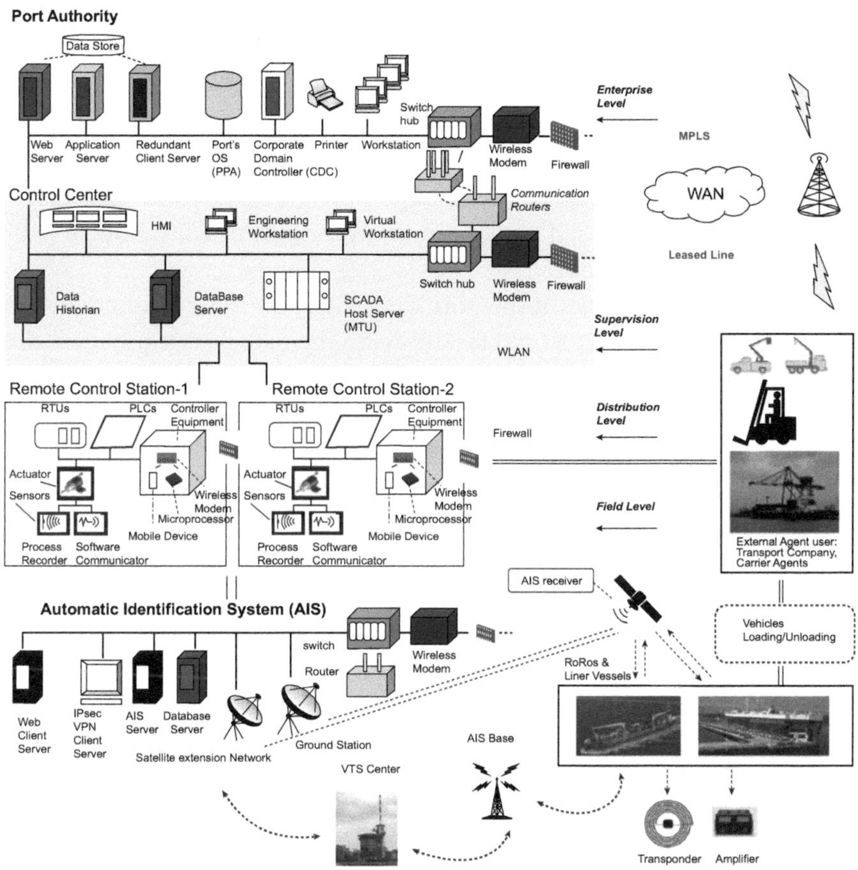

Figure 2.4 Vessel arrival and vehicles uploading process.

processes through a single request submission. In this context, the PCS checks the custom clearance document authentication by data verification services requesting information from the ICIS via an Electronic Data Interchange system and reports the result. It should be noted that during the clearance arrangement and docking authorization processes, information on vessels position and an arrival/departure timetable is obtained via the e-tracking (MarineTraffic Internet Automatic Identification System [AIS]) services hosted by a "MarineTraffic" web platform. These kinds of services are operated via satellite network communication systems, supported by an AIS that generally consist of satellite devices, antennas, modem, routers, power stations, satellite terminals, bandwidth options, IP Network, local intranet, and remote-control systems.

Then, once all relevant formalities have been arranged by the authorized entities (e.g., docking clearance submission, operations through Supervisory Control and Data Acquisition [SCADA] systems) and the port's equipment (yard tractors, forklifts with the auxiliary equipment like sensors, etc.) and port personnel are available to unload the vehicles and store them to the port's facilities, the ship agent informs the ship about the docking arrangements using software-based satellite communications or conventional radio communications systems.

When the vessel arrives at the port, the vehicle unloading from the vessel is performed by yard tractors and forklifts with their auxiliary equipment that are monitored by an SCADA system. This system is a common process automation system that is used to gather data from sensors and instruments located at remote points and to transmit them at a central point for either control or monitoring purposes. In this way, the system can analyze and display information improving the performance of vehicles shipping operations, reducing waste of time and providing cost saving. To meet its objectives, it incorporates a bundle of components such as the following: (1) a human machine interface (HMI) that is an input–output device acting as a central point of human monitoring and control of the vehicle unloading processes, storing databases, and display of statistical control charts, and reports; (2) programming logic controllers (PLCs) are solid-state computers reading signals from devices such as sensors, switches, and keyboards capable of monitoring the motion of objects and automating harsh industrial processes by providing electromechanical operations during vehicle loading and unloading of the vessels; (3) remote terminal unit (RTU) devices convert electrical to digital signal, feeding telemetry data to the central monitoring station of an SCADA system; RTUs are considered microprocessor electronic devices connecting physical equipment with the supervisory stations via WLAN and receive messages from the master station to control the interconnected objects; (4) an SCADA master terminal unit (MTU), which is, typically, a real-time data repository; (5) a data historian database for storing data commands; (6) supervisory stations that are responsible for communicating between SCADA equipment such as RTUs or PLCs and HMI software for data acquisition; and (7) trailer cranes and forklifts for vehicles unloading from the vessel, which are also monitored and controlled via sensors of the SCADA system. All communications between the aforementioned SCADA components are performed using wired and/or wireless network connections.

Various threats may occur in the ports' SCSs. In Table 2.2, we show possible threat categories that may occur during the vehicles' transport from an

Table 2.2 Threat analysis of the SCS "Vehicles' Transport"

Use case entities	Possible threat categories
Industry	Destroy critical SC infrastructure Unauthorized access to SC infrastructures Information and ICT threats Personnel security and safety threats Goods and conveyance security threats
Move to port	Information tampering Information loss Personnel security and safety threats Goods and conveyance security threats
Load cargo ship	Destroy critical SC infrastructure Unauthorized access to SC infrastructures Information and ICT threats Personnel security and safety threats Goods and conveyance security threats
Travel to import port	Destroy critical SC infrastructure Unauthorized access to SC infrastructures Information and ICT threats Personnel security and safety threats
Unload from cargo ship	Destroy critical SC infrastructure Unauthorized access to SC infrastructures Information and ICT threats Personnel security and safety threats Goods and conveyance security threats
Custom authorities	Destroy critical SC infrastructure Unauthorized access to SC infrastructures Information and ICT threats Personnel security and safety threats Goods and conveyance security threats
Transport to importer	Information tampering Information loss Personnel security and safety threats Goods and conveyance security threats
Importer warehouse	Destroy critical SC infrastructure Unauthorized access to SC infrastructures Information and ICT threats Personnel security and safety threats
Importer	Destroy critical SC infrastructure Unauthorized access to SC infrastructures Personnel security and safety threats SC as a means of smuggling Unauthorized use

Continued

Table 2.2 Threat analysis of the SCS "Vehicles' Transport"—cont'd

Use case entities	Possible threat categories
Negotiate contract with industry	Personnel security and safety threats SC as a means of smuggling Unauthorized use
Negotiate contract with local agent	Personnel security and safety threats SC as a means of smuggling Unauthorized use
Negotiate contract with ship agent	Personnel security and safety threats SC as a means of smuggling
Negotiate contract with ship owner	Personnel security and safety threats SC as a means of smuggling
Prepare documentation for customs	Destroy critical SC infrastructure Unauthorized access to SC infrastructures
Cargo processed by customs	Personnel security and safety threats

industry until the receipt of the vehicles by the *importer*. Note that green entities are related to the cargo flow (physical layer) while red entities are related the flow of information and documents (cyber layer).

Let us also see two specific attack scenarios enabling the stealing of vehicles in the "Vehicle Transport" SCS:

Attack Scenario I: Remote Access

We assume that a criminal gang wants to steal vehicles from the vehicles terminal of a port. Toward that, hired hackers can engage into malicious activities spanning from simple phishing attacks (targeting port authorities and key employees) to the exploitation of more sophisticated, remote malware targeting the onboard communication interfaces and units of the vehicles in question. In the first case, the hackers begin to survey the port authority by looking through public documentation about the company, such as its web page and relevant news articles. After correlating this information and identifying several key employees, which are suspected to have partial access to relevant systems at the port, they then compose and send specially crafted emails with content relevant to them (possible extracted from social media and other similar information sources). Each email includes a PDF document that from a user's perspective contains some interesting data. However, opening this PDF triggers the execution of an attached exploit (for a publicly known vulnerability in Adobe Reader) that silently installs a remote access service on the computer [71].

The criminals manage to compromise a few computers and critical elements of the network by exploiting various such vulnerabilities, one of which is especially interesting as the user is a domain administrator with elevated privileges to most IT systems at port. The criminals use the credentials of the administrator to gain remote access to the port system. Here, they identify the precise location of the relevant vehicles within the port and some especially valuable truck-based transports that have been scheduled to be carried out. Then the hackers changed their location and delivery times, and the gang then sent in their own drivers to collect the vehicles ahead of the scheduled pick up.

On the other hand, in the second case, the hackers managed to get access to more sophisticated attack tools capable of identifying and exploiting vulnerabilities that pertain to the in-vehicle communication interfaces, e.g., mobile communications, near field communications, wireless sensor networks, etc. By launching attacks based on software-related vulnerabilities, dynamic memory errors, etc., they manage to get access to the vehicle's vast network of interconnected onboard units and eventually spoof their geolocation. Examining the in-port vehicle scheduling processes followed, the criminals can then change the route and location of the vehicles, to their preferred points of interest, without the port system administrator detecting any of these changes.

In both scenarios, the criminals managed to disrupt the functionality of a subset of port system functionalities for stealing targeted vehicles. In addition, the hackers could exploit vulnerabilities in the surveillance system of the port that controls the CCTV video cameras to gain access and delete video streams that show their malicious activities.

Attack Scenario II: SCADA Attack

The attack scenario falls into the following description:

> *The criminals manage to compromise some critical elements of the SCADA system by exploiting various vulnerabilities, one of which is especially interesting as the MTU user is a domain administrator with elevated privileges to most SCADA assets at the port. The criminals use the credentials of the MTU administrator to gain remote access to the SCADA system, and by exploiting the vulnerability that allows them to execute code, they penetrate the HMI system. Consequently, they identify the precise location of the relevant vehicles within the port and some especially valuable truck-based transports. Then, the hackers send their location and delivery times to their own drivers and change the relevant data to collect the vehicles.*

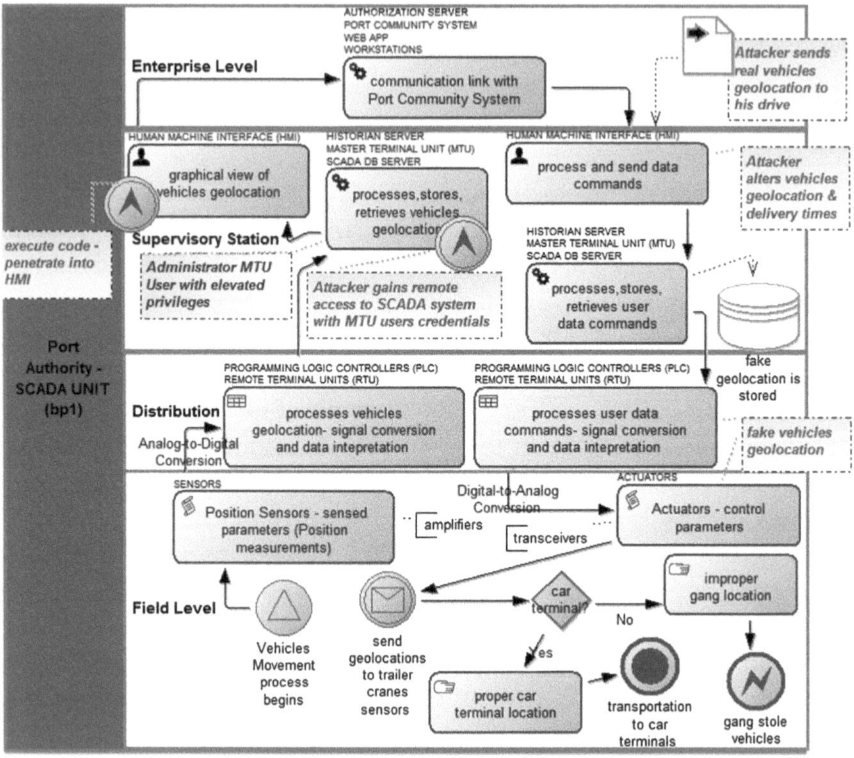

Figure 2.5 Execution steps of the "vehicles robbery" attack during vessels unloading.

In this attack scenario, criminals, because of the limited security controls, penetrate the HMI system and alter the geolocation of vehicles according to their mean purposes. To achieve this, they exploit two types of vulnerabilities, found in HMI hardware: the gain privileges and code execution types. Fig. 2.5 shows the BPMN model of this attack scenario. The attacker's performance over the process is depicted with the event BPMN elements. More specifically, gaining the MTU administrator privileges and executing code to penetrate the HMI system is shown with the escalation (circle) event element of BPMN 2.0. The potential of such an attack generates alternative paths in the process, initialized from the exclusive gateway diamond element of BPMN 2.0 after the vehicles delivery event message (Fig. 2.5): (1) the normal path, where criminals have not performed the attack on SCADA assets, ends up with the vehicles transported to a car terminal (shown with the terminate end event of BPMN 2.0 in Fig. 2.5); (2) the improper path where criminals have performed the attack and process

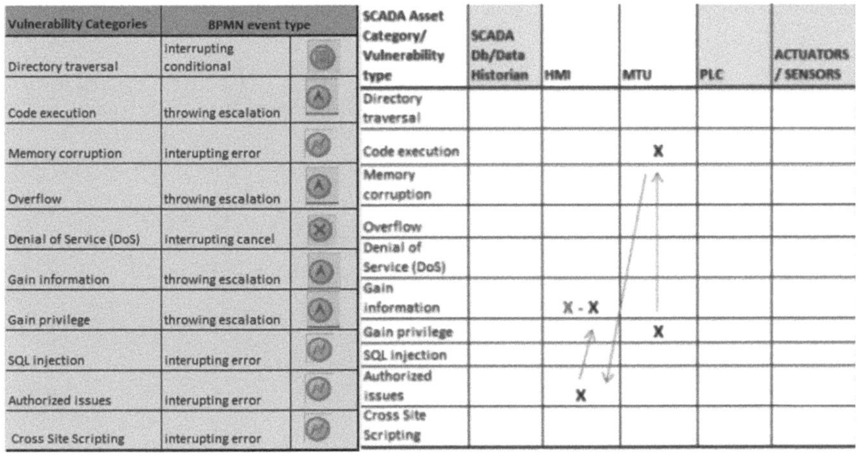

Vulnerability Categories	BPMN event type
Directory traversal	interrupting conditional
Code execution	throwing escalation
Memory corruption	interupting error
Overflow	throwing escalation
Denial of Service (DoS)	interrupting cancel
Gain information	throwing escalation
Gain privilege	throwing escalation
SQL injection	interupting error
Authorized issues	interupting error
Cross Site Scripting	interupting error

SCADA Asset Category/ Vulnerability type	SCADA Db/Data Historian	HMI	MTU	PLC	ACTUATORS / SENSORS
Directory traversal					
Code execution			X		
Memory corruption					
Overflow					
Denial of Service (DoS)					
Gain information		X - X			
Gain privilege			X		
SQL injection					
Authorized issues		X			
Cross Site Scripting					

Figure 2.6 Left image: mapping SCADA vulnerabilities with BPMN events. Right image: vulnerabilities of SCADA asset.

closes with the gang committing the robbery of the vehicles (shown with the error end event of BPMN 2.0, also depicted in Fig. 2.5).

In the described attack scenario, attackers exploit two types of vulnerabilities: gain privileges, and execute code, which are both presented in the BPMN model with the throwing escalation event of BPMN 2.0. This is because both vulnerabilities reach an escalation to be exploited. Likewise, we can express some important types of vulnerabilities that are possible to appear in SCADA assets with the boundary event elements of BPMN 2.0, namely, events that appear in boundaries of BPMN task elements, as shown in Fig. 2.6. For example, the vulnerability type "directory traversal" is expressed with the interrupting conditional.intermediate event, as it declares the condition that a part of the application is vulnerable. Alike, a "denial of service" is characterized with the interrupting cancel event as in this case the user is deprived of the services. Vulnerability types such as "memory corruption" or "SQL injection" are presented with the interrupting error event as they generate errors.

SCADA systems are critical components of current industrial automation systems underlying high security risks. SCADA systems are mainly used in ports' CIs and their maritime logistics and SCSs providing cost reduction and saving time in their operations and processes. As cyberattacks rapidly increase in the maritime ecosystem, SCADA security needs to be highly considered.

CHAPTER 3

Security of Ports' Critical Information Infrastructures

The main concern of all organizations is to be able to identify their threats and estimate their risks, which is the main goal of risk management, i.e., to protect business assets (physical and cyber) and minimize costs in case of failures. Security management represents a core duty of successful corporate governance. Hence, risk management describes a key tool for the security within organizations, and it is essentially based on the experience and knowledge of best practice methods. These methods consist of an estimation of the risk situation based on the business process models and the infrastructure within the organization. In this context, these models support the identification of potential risks and the development of appropriate protective measures. The major focus lies on the organizations (e.g., port authorities) for the identification, analysis, and evaluation of threats to the respective corporate values.

The outcome of a risk analysis is in most cases a list of risks or threats to a system, together with the corresponding probabilities. International standards in the field of risk management are used to support the identification of these risks or threats as well as to assess their respective probabilities. These standards range from general considerations and guidelines for risk management processes to specific guidelines for the IT sector all the way to highly specific frameworks as, for example, in the maritime sector. Most of these standards specify framework conditions for the risk management process but rarely go into detail on specific methods for the risk analysis or risk assessment. This is one reason why differences in the risk assessment often arise within the specific areas of application, making a direct comparison of the results difficult. Furthermore, the aforementioned efforts are not sector specific; as a result, they are too generic and difficult to be applied in the complex maritime sector.

In principle, choosing the right method and the right tool for risk analysis and risk evaluation proves to be complicated. A huge emphasis in the security and risk management in the maritime sector is laid on the physical security. The International Ship and Port Facility Security (ISPS) Code (as well as the respective EU regulation) defines a set of measures to enhance

Port Cybersecurity
ISBN 978-0-12-811818-4
http://dx.doi.org/10.1016/B978-0-12-811818-4.00003-4

the security of port facilities and ships. Additionally, several methodologies and tools exist aimed at strengthening the safety level of the ports' infrastructures (physical risk assessment). Nevertheless, due to the increased interaction and exchange of port's information with other critical infrastructures in the maritime ecosystem (e.g., port authorities, ministries, maritime companies, ship industry, etc.) the sole focus on physical security is not sufficient anymore. In the same way, the security of the ports' ICT and physical-related components, elements, and systems becomes equally important.

However, by security management, we mean the effective implementation, establishment, assessment, monitoring, improvement, and auditing of the security of the ICT system (all assets in the six layers of the ports' ICT systems). Managing security requires a continuous and systematic process of identifying, analyzing, mitigating, reporting, and monitoring technical, operational, and other types of security risks (risk management) as well as implementing appropriate security measures and controls. Although various efforts and processes can be found in security management of ports' critical infrastructures (CIs), none of them address the security management of all layers of ports' CIs. They only treat the physical security, and they only deal with safety management of port systems.

SAFETY MANAGEMENT: A RESTRICTING APPROACH

Traditionally, targeted methodologies for risk assessment of ports like MSRAM (Maritime Security Risk Analysis Model) and its extended version MSRAM-PLUS/FORETELL address only physical security, and they are only compatible with the ISPS. Similarly, the available maritime risk assessment systems like MARISA concentrate on the safe navigation of ships during their presence in the port. The risk assessment system CMA detects abnormal behavior of ships and identifies respecting physical threats.

The ILO Port Health and Safety systems recognizes that risk assessment is an essential part of safety management. It provides a sound basis for the improvement of safety. It covers tasks and physical hazards in the workplace and allows hazards to be assessed to see how harmful they are. But cybersecurity threats are not covered in the ILO code.

The World Bank Group, together with the International Finance Corporation, released in 2007 a document called "Environmental, Health, and Safety Guidelines for Ports, Harbours, and Terminals." The Environmental, Health, and Safety Guidelines are technical reference documents with general and industry-specific examples of Good International Industry Practice concentrating on physical hazards.

Various research efforts concentrate on physical threats, ignoring the cyber risks. Some examples follow.

Safety4Sea is a dedicated Maritime Safety and Environmental portal, a PRO BONO project to promote maritime safety and environmental awareness, operational safety, and environmental excellence. Safety4Sea's mission is to make practical safety and environmental excellence easy to understand for everyone in the industry, promote best practices, and improve people perception by promoting safety and environmental awareness in a wide range of maritime aspects.

FLAGSHIP is a partially EU-funded project, focusing on improvement of safety, environmental friendliness, and competitiveness of European maritime transport. The project contributes to a further increase in the capacity and reliability of freight and passenger services and to a reduction of negative impact from accidents and emissions. The emphasis of the project is on onboard systems and procedures, ship management systems on shore, impact of new technology on present ship owner and operator organizations, effective and efficient communication interfaces, and impact of standards and regulations. FLAGSHIP aimed to create the mechanism by which the expertise of all the required actors can be brought together in real time, independently of their location, and given to the right people, in the right format, at the right time, and incorporating the highest level of knowledge, so they can better manage all the questions that confront a ship operator: issues relating to the ship itself and its equipment (e.g., hull monitoring, equipment diagnostics, maintenance planning), its day-to-day operation (e.g., navigation, cargo, rule compliance), as well as emergencies and other exceptional situations (collision, fire, etc.).

The SafePort was a collaborative project under the EU Seventh Framework Programme. Many European ports will reach full capacity in the next few years. SafePort takes its cue from the aviation industry, which has addressed safety issues created by increased traffic through increasing automation and the use of sophisticated traffic management systems. SafePort developed and demonstrated an active vessel traffic management and information system (A-VTMIS) to manage vessel movement within its jurisdiction. This will ensure that vessels follow safe paths without conflicting with other vessels and improve the efficiency of port operations.

Several Preparatory Action on Security Research and EU FP7 Security Research programs initiated over the last decade addressing issues relating to strengthening the safety of ports and/or their ports' CI systems. Most of these projects have fallen in three main categories, as follows:

Improved maritime surveillance systems: by enhancing the interoperability of local and national surveillance systems through the pooling of cross-sectoral surveillance information and its fusion into a central database. Representative examples are the following:

- The Autonomous Maritime Surveillance System (AMASS) project focused on strengthening maritime surveillance and on better integrating information and data between relevant agencies. The focus was on developing a cutting-edge early warning system that provides maritime authorities and law enforcement agencies with information about attempts at illegal immigration and other criminal activities at sea.
- The Underwater Coastal Sea Surveyor project is a cost-effective response to new terrorism attacks especially against underwater improvised explosive device threats. It provides a fundamental technology for the global issue of maritime surveillance and port/naval infrastructure protection.
- The Surveillance of Borders, Coastlines and Harbors project attempted to combine and maximize the use of existing surveillance technologies to model the most effective operational procedures for enhancing the surveillance of borders, coastlines, and harbors.
- The Sea Border Surveillance (SEABILLA) project aims to define the architecture for cost-effective European sea border surveillance systems, integrating space, land, sea, and air assets, including legacy systems. The project is applying advanced technological solutions to improve the performance of surveillance functions.

Interoperability of ports' CI systems: by enhancing the capability to collect and merge maritime-related data into a common and comprehensive picture to be shared among relevant organizations. Projects of this category are the following:

- The InterOPERAble Approach to European Union MARitime Security Management (OPERAMAR) project attempted to solve the issue of fragmentation between member states caused by the persistence of nation-specific procedures, legislations, and systems that hamper interoperability, greater information sharing, and improved coordination.
- The SECure CONtainer Data Device (SECCONDD) project was designed to initiate the international standardization of the technical interface between a secure container or vehicle and a data reader at a port or border crossing. The interface should enable law enforcement

and trade officials to read security data, including stored information from internal security and location sensors.

Protection of critical maritime infrastructure: by mitigating the risks of maritime safety (physical) incidents. A notable number of projects are the following:

- The Security System for Maritime Infrastructure, Ports, and Coastal Zones (SECTRONIC) project attempted to improve the safety of civilian ships (passenger and cargo carriers), energy platforms and facilities, and ports through advanced information, sensor, and response technologies. It aimed to develop an integrated security system combining surveillance, intrusion detection, and response to events and incidents.
- The Security Upgrade for Ports (SUPPORT) project aims to raise the current level of port safety by integrating legacy port systems with new surveillance and information management systems. Furthermore, the SUPPORT project has a special focus on border control, aiming to secure uninterrupted flows of cargos and passengers while allowing for the effective elimination of illegal immigration and trafficking.

CYBERSECURITY REGULATIONS AND STANDARDS

Most ports are compliant with ISPS code; however, this compliance does not imply secure ports since ISPS only addresses organizational and safety issues. However, the most recent regulations and directives are cybersecurity focused, and their implementation at the EU and international level will help the port authorities to secure their ICT systems and to better mitigate existing and upcoming cyber risks. In the upcoming years, the ports will need to implement the following new regulatory framework:

- **CIIP Directive** (2012), critical information infrastructure protection: toward global cybersecurity;
- **The Cybersecurity Strategy for the European Union** (2013) and the **European Agenda on Security** (2015) provide the overall strategic framework for the EU initiatives on cybersecurity and cybercrime;
- **eIDAS Regulation** (2014) on electronic identification and trust services for electronic transactions in the internal market;
- **European Parliament** (2015) concerning measures to ensure a high common level of network and information security across the union;
- **NIS Directive** (2016) applies only to those public administrations that are identified as operators of essential services;

- **cPPP Initiative** (2015) ensures that Europe will have a dynamic, efficient, and effective market in cybersecurity products and services;
- **Enhanced Privacy Directive** (2016), mandatory reporting of security breaches;
- **USA H.R. 3878**, House of Representatives, "Strengthening Cybersecurity Information Sharing and Coordination in Our Ports Act of 2015."

Besides the new upcoming regulatory framework, standards and best practices embrace the effort for better ICT security management. An overview of ICT security management standards is presented in this section. It should be noted that these standards do not constitute risk management methods, but rather, they fix a minimal framework and describe requirements, for the risk assessment process itself, for the identification of the threats and vulnerabilities allowing to estimate the risks, their level, and then to be able to define an effective treatment plan.

The most well-known cybersecurity standards are these:

ISO/IEC 27001 36 is a commercial standard that specifies requirements for the establishment, implementation, monitoring and review, maintenance, and improvement of an information security management system (ISMS). The ISMS is an overall management and control framework for managing an organization's information security risks. The ISO/IEC 27001 does not mandate specific information security controls but stops at the management and operational level. Usually, a group of analysts with high ICT expertise and experience verifies the compliance of the organization with the defined requirements. However, although the compliance process requires the involvement of multiple users, the collaborative abilities of the standard are limited due to its inherent complexity. The standard covers mostly large-scale organizations (e.g., governmental agencies and large companies), while it is considered too heavy for micro, small, and medium size businesses.

The ISO/IEC 27001 ISMS incorporates several Plan-Do-Check-Act cycles: for example, information security controls are not merely specified and implemented as a one-off activity but are continually reviewed and adjusted to take account of changes in the security threats, vulnerabilities, and impacts of information security failures, using review and improvement activities specified within the management system. There exist a variety of freeware (e.g., EBIOS developed by Central Information Systems Security Division [France]) and commercial software (e.g., CRAMM developed by Insight Consulting) that verify the compliance of the organization with the ISO/IEC 27001.

ISO/IEC 27005:2008 38, a commercial standard from the International Organization for Standardization (ISO) and the International Electrotechnical Commission (IEC), describes the risk management process and its activities for information security and provides guidelines for information security risk management and supports the general concepts specified in ISO/IEC 27001:2005 as well as the main principles and rules described in ISO/IEC 27002:2005. It is applicable to all types of organizations (e.g., governmental agencies, large companies, small and medium size enterprises) that intend to manage cyber risks that could compromise the organization's information security. Essentially, the ISO information security risk management process can be applied to the whole organization; any discrete part of the organization (e.g., a department, a physical location, a service); any ICT system; and any existing, planned, or aspect of control (e.g., business continuity planning).

ISO 27005 proposes the use of both quantitative and qualitative methods for the calculation of the risk level; however, it does not support any specific technique for this purpose or any computational method to analyze and combine the assessment information. The generic nature of the standard does not include aspects that promote the collaboration among the users.

In this context, more integrated risk management methodologies and methods such as EBIOS, MAGERIT, and MEHARI comply with the rules and obligations defined by the specific standard.

ISO/IEC 27002:2005 is a commercial standard that establishes guidelines and general principles for initiating, implementing, maintaining, and improving information security management in an organization. It provides specifications with guidance for implementation of the ISMS in the organization. This can be used by internal and external analysts with high ICT expertise and experience, to assess an organization's ability to meet its own requirements, as well as any customer or regulatory demands.

The standard provides a list of 10 main control domains (organization of information security; asset management; human resources security; physical and environmental security; communications and operations management; access control; information systems acquisition, development, and maintenance; information security incident management; business continuity management; compliance) comprising 36 control objectives and 127 controls, which are used for the assessment. The standard promotes the adoption of a process approach for establishing, implementing, operating, monitoring, and improving the effectiveness of an organization's ISMS.

It should be noted that ISO/IEC 27002 is not a real method for risk analysis and management, but rather compliance standards, reporting a list of controls for good security practices and the requisites that an existing method should have to be standard compliant. However, although it is neither a method for evaluation nor for management of risks, it includes specific risk handling aspects such as the identification of risk and the creation of an initial risk treatment plan. The standard can cover all types of organizations (e.g., governmental agencies) and all sizes from micro to medium and large size businesses.

SECURITY MANAGEMENT: A HOLISTIC APPROACH

Since an ICT system as defined in Chapter 2 is a six-layer system comprised by physical and cyber layers and assets, a holistic approach to risk assessment is needed to assess the security (confidentiality, integrity, authenticity, availability) level of all assets (physical and cyber) in all layers of the ports' ICT systems. In this section, an overview of security-related legislation and ICT security management standards are provided.

According to the survey [160] the existing methodologies use *four different approaches* to define the risks:

1. The concept of the risk is related to a threat and an asset (or a group of assets), and it comprises the likelihood of the threat, the vulnerability level of the asset(s) to the threat, and the impact of the threat on the asset(s).

$$\text{Risk (Threat, Asset)} = \text{Likelihood (Threat)} \otimes \text{Vulnerability (Threat, Asset)} \otimes \text{Impact (Threat, Asset)}$$

2. The concept of the risk is related to a threat, an asset, and specific security needs. It comprises the vulnerability of the asset and the impact of the threat on the security needs.

$$\text{Risk (Threat, Asset, Needs)} = \text{Impact (Threat, Needs)} \otimes \text{Vulnerability (Threat, Asset)}$$

3. The concept of the risk (defined as annual loss expectancy [ALE]) is related to a threat and an asset, and it comprises the probability of the threat affecting the asset and the average loss of the resulting incident.

$$\begin{aligned}\text{Risk (Threat, Asset)} &= \text{ALE (Threat, Asset)} \\ &= \text{Probability (Threat, Asset)} \\ &\quad \otimes \text{Average Loss (Threat, Asset)}\end{aligned}$$

4. The concept of the risk is related to a threat and a critical asset, and it comprises the impact of the threat on the critical asset and the vulnerability of the asset.

$$\text{Risk (Threat, Critical Asset)} = \text{Impact (Threat, Critical Asset)} \otimes \text{Vulnerability (Critical Asset)}$$

5. The concept of the risk is related to an incident (i.e., a threat exploiting vulnerability) and an asset, and it comprises the likelihood of the incident and the consequences of the incident itself.

$$\text{Risk (Incident, Asset)} = \text{Likelihood (Incident)} \otimes \text{Consequences (Incident, Asset)}$$

The **security management process**, i.e., the process in evaluating all risks, consists of the following phases:

- *context establishment*: intends to define the risk management's boundary (e.g., the entire organization, one ICT, one service);
- *asset identification*: identify all assets within the boundary of the assessment;
- *threat analysis*: analyze all threats of the identified assets and the threat levels;
- *risk analysis*: intends to evaluate the risk levels;
- *risk assessment*: used to make decisions and consider the objectives of the organization;
- *risk treatment*: to treat, reduce, retain, avoid, or transfer the risks;
- *risk communication*: to achieve agreement on how to manage risks by exchanging and/or sharing information about risk between the decisionmakers and other stakeholders;
- *risk monitoring and review*: to detect any chances in the context of the organization at an early stage, and to maintain an overview of the complete risk snapshot.

The existing security management methodologies can be evaluated based on the following *assessment criteria* as proposed by the literature [160,33]:

C1. *scope*: the applicability of the method. The following types have been identified: (1) general-purpose method, covering only specialized ICT requirements and (2) targeted-purpose method, covering specific sectoral characteristics, particularities, needs, and requirements;

C2. *target group*: the most appropriate type of organizations the method aims at;

C3. *RA/RM support*: the phases that the method supports (risk analysis or/and risk management);
C4. *evaluation scale*: the approach (quantitative or qualitative) used in the methods to evaluate the risk level;
C5. *impact evaluation*: the approach adopted in the methods to determine the impact level. Each method uses specific scenarios, factors, parameters, and guidelines to define the impact of an event;
C6. *risk evaluation*: the approach used in the methods to calculate risk level;
C7. *collaboration capabilities*: the capacity of the methods to promote the collaboration of the users in the evaluation process;
C8. *computational capabilities*: the capacity of the methods to analyze and combine diverse and distributed corporate knowledge;
C9. *required skills*: the level of skills needed to use and maintain the method;
C10. *cost*: the licensing schema available for the method;
C11. *automated tools*: availability of tools that support the method;
C12. *compliant with standards*: compliance with national or international standards.

The existing security management methodologies are assessed for their suitability to be applied in the ports' ICT CIIs. Because commercial ports are CIs, we also present risk assessment methodologies for CIs. This section concludes that methodologies combining maritime safety, cyber, and CI standards will be most appropriate for the security management of commercial ports CIIs.

For an enriched repository of security risk management methodologies and tools, we refer the reader to the ENISA repository [67]; in this section, we describe only the most well-known ones:

NIST 800-30 [94] is a free guide that provides a foundation for the development of an effective risk management program, containing both the definitions and the practical guidance necessary for assessing and mitigating risks identified within IT systems. The goal is to help mostly large-scale organizations (such as governmental agencies and large companies) to better manage IT-related mission risks. The use of a primitive method for the calculation of the risk level in combination with the lack of an effective computational technique to analyze and correlate the knowledge located in a corporate environment reduce the abilities of the NIST 800-30 for a more integrated approach. In addition, even though the method has adopted and uses extensive technical and operational questionnaires that require the involvement of a variety of users, the concept of collaboration in the determination of the overall results and the formulation of the final treatment plan is limited. The risk analysis and management process defined in NIST 800-30 is usually executed by a dedicated group of ICT experts.

The method is compliant with the ISO/IEC 27001:2005 addressing all the requirements for the establishment and implementation of an ISMS. Currently, NIST 800-30 is not supported by any freeware or commercial application.

OCTAVE [115] is a free of charge approach to information security risk evaluations that is comprehensive, systematic, context-driven, and self-directed. The approach is embodied in a set of criteria that define the essential elements of an asset-driven information security risk evaluation. Initially, it was designed with larger organizations in mind; however, a targeted method for small organizations has been developed.

The OCTAVE method uses a three-phase approach to examine organizational and technology issues, assembling a comprehensive picture of the organization's information security needs. The method uses workshops to encourage open discussion and exchange of information about assets, security practices, and strategies. In this context, the corporate users participate actively in several parts of the evaluation process. This indicates that the method incorporates specific collaborative abilities.

Each phase consists of several processes, and each process has one or more workshops led or conducted by the analysis team. Some preparation activities are also necessary to establish a good foundation for successfully completing the evaluation. OCTAVE uses for the risk analysis a primitive approach based on a qualitative scale (high, medium, low). In addition, the method does not integrate an advanced technique for the analysis and combination of the knowledge located in the corporate environment. Thus, the information is going through an insufficient process for the determination of the overall results. Finally, OCTAVE method is supported by commercial standalone software, OCTAVE Automated Tool, implemented by Advanced Technology Institute. The tool can assist the user during the data collection phase, organizes collected information, and finally, produces the study reports.

CRAMM [55] is a method developed to assist mostly the large-sized organizations (such as governmental agencies and large companies) to undertake a risk analysis of information systems and networks, to identify security requirements and possible solutions, and to detect contingency requirements and possible solutions. The method is applicable to all types of information systems and networks and can be applied at all stages in the information system lifecycle, from planning and feasibility, through development and implementation, to live operation.

CRAMM consists of three main phases: identification and valuation of assets, risk analysis, and risk management. In this method, an analyst or a group of analysts undertake the responsibility to evaluate the security and risk level of the organization analyzing and combining the diverse knowledge distributed in the corporate environment. The computational method and technique that has been adopted by CRAMM for the correlation and the determination of the results is quite primitive and is based on a qualitative approach. In addition, the involvement of the users of the organizations to the actual assessment can be considered low; thus the collaborative capabilities of the method are characterized as limited. For the analysts to use and execute all the phases of the method (identification and valuation of the assets and risk analysis and management), they should have a high level of skills and experience at gathering and analyzing information to identify threats and vulnerabilities, to infer the risks, and to define the most appropriate countermeasures that fit to the needs of the organizations.

As a method, CRAMM is detailed enough and can cover an extensive range of features at management, operational, and technical levels. Also, CRAMM complies with the rules and obligation imposed by the ISO 27001 and ISO 27005 standards. CRAMM is supported by a commercial standalone tool, developed by Insight Consulting, that provides a way to implement the proposed method.

EBIOS [42] is a risk management approach created under the French General Secrétariat of National Defense. It proposes a methodology and supporting software for assessing and treating risks in the field of information systems security.

EBIOS approach consists of a cycle of the following phases: context analysis in terms of global business process dependency on the information system; threat analysis; and risks estimations. EBIOS methodology is easy to understand and deploy; thus it can be applied by a set of organizations that vary from governmental agencies and large companies to small and medium size enterprises. Its overall philosophy is straightforward and intuitive, and it follows a natural sequence. It consists of formalizing the sensitivities and threats and determining the associated risks for the organization. The methodology possesses collaborative abilities since it gathers and combines the corporate knowledge in a smooth and efficient manner based on a qualitative approach. However, the lack of an advanced computational schema for the correlation and determination of the results can be considered a main disadvantage.

EBIOS has been applied both to basic systems and to complex systems (human resources management system interconnecting several elements), at the predesign stage or on existing systems, to complete information systems or to subsystems. Although, it should be noted that the level of detail of the method is limited to management and operational issues and characteristics.

EBIOS can cover all the requirements, steps, and processes defined by a variety of IT standards such as the ISO/IEC 27001:2005, the ISO/IEC 27002:2005, and the ISO/IEC 27005:2008. The method is supported by an open source tool developed by Central Information Systems Security Division (France) and is a standalone application that is based on Java and XML technologies. The tool integrates all risk analysis and management steps defined by the five EBIOS phases assisting users with low IT expertise and experience to evaluate and mitigate the corporate risks.

IT-Grundschutz [11–13,26] has been developed by the Federal Office for Information Security in Germany, and it provides a configuration for the establishment of an integrated and effective ICT security management.

The method, before starting the risk analysis, does a basic security check to verify implemented security measures. Risk assessment identifies threats, which are not avoided by the measures, such as residual threats. These threats can be eliminated by additional security measures. In this way, risk will be reduced to an acceptable level.

IT-Grundschutz has been designed to apply to organizations with complex underlying infrastructure such as governmental agencies and large companies as well as to small and medium size businesses with basic systems. The method can be deployed by users with standard IT-related expertise and experience that undertake the responsibility to execute the evaluation process. However, the collaborative abilities of the method can be considered low since the corporate users are involved only in specific steps of the risk assessment. The method is compliant to ISO/IEC 27001:2005, addressing the defined requirements, as well as being suitable for the implementation of the ISMS process described by ISO/IEC 27002:2005. In this context, IT-Grundschutz aims at assisting all users with managerial, operational, and technical responsibilities in their efforts to manage the security of information and ICT resources and to reduce the associated risks.

The method is supported by commercial software, GStool, developed by Federal Office for Information Security (BSI). GStool is a standalone application with database support.

MAGERIT is an open methodology developed by the Spanish Higher Council for Electronic Government, offered as a framework and guide to the public administration. It is the answer in the increasing dependency of the public and private organizations on information technologies to fulfill their mission and reach their business objectives. The purpose of MAGERIT is directly related to the generalized use of ICT systems that bring evident benefits for the users but which is also subject to certain risks that must be kept under control by means of security countermeasures that generate confidence in the use of these media.

Various organizations that possess complex IT infrastructure (governmental agencies and large companies) as well as basic systems (small and medium size enterprises) can apply this method to identify and mitigate their security risks. MAGERIT can be used and maintained only by users with high ICT expertise and experience. These users undertake the responsibility to run the risk analysis process via workshops and interviews with specific representatives of the organization that participate only in specific phases of the assessment process. In this context, the method does not support sufficient collaborative capabilities and features.

MAGERIT complies with a set of IT standards. Specifically, it addresses all the rules and obligations imposed by the risk analysis and management standards ISO/IEC 27005:2008, covers all the requirements defined by the ISO/IEC 27001:2005, and conforms with the code of implementation of an ISMS specified by the ISO/IEC 27002:2005.

A commercial software that implements and expands the MAGERIT methodology is the EAR/PILAR. This is a standalone application (based on Java and XML technologies), developed by A. L. H. J. Mañas that has been designed to support and execute the defined risk management process.

MEHARI is a free of charge qualitative risk analysis and management method developed by CLUSIF (CLub for the Security of Information in France or CLub de la Sécurité dde l'Information Français). MEHARI provides a consistent methodology, with appropriate knowledge bases (e.g., manuals and guides that describe the different modules [stakes, risks, vulnerabilities]), that has been designed to assist people implicated in security management (CISOs, risk managers, auditors, CIOs) in their different tasks and actions. Specifically, it is targeted to users with managerial, operational, as well as more technical responsibilities. The methodology is suitable for the implementation of the ISMS process described by ISO/IEC 27001:2005; it is compliant with ISO/IEC 27005:2008 requirements providing the set of tools and elements required for its implementation.

MEHARI is most appropriate for medium- to large-scale organizations such as governmental agencies and medium and large size companies. The corporate users can participate only in specific phases of the methodology related to the identification of assets and vulnerabilities. In this context, the collaborative capabilities of the methods can be considered limited, since the users are not involved directly in the risk calculation and the formulation of the risk treatment plan. In addition, the method uses a primitive computational method to analyze and combine the diversity of the information to deduce the final results.

MEHARI is supported by two standalone toolkits. The first one is commercial software managed by the company Risicare, and the second is a freeware application, [MEHARI2010] basic tool, developed by CLUSIF.

ISAMM [54] is a quantitative type of risk management methodology that can be applied by a variety of organizations such as governmental agencies, large companies, and small and medium size enterprises. In this method, the assessed risks are expressed, through their ALE, in monetary units. ALE is the annual expected loss or cost should a threat or a group of threats be materialized.

$$\text{ALE} = [\text{probability}] \times [\text{average impact}]$$

This formulates the basis for the return on investment–based approach and the economic justification capabilities of ISAMM with respect to the risk treatment plan. ISAMM allows showing and simulating the reducing effect on the risk ALE for each improvement control and to compare this with its cost of implementation.

ISAMM is compliant to ISO/IEC 27002 and provides maximal support of the ISO/IEC 27001 ISMS standard. It is supported by a freeware tool, ISAMM Consultant tool, and a commercial application named ISAMM Client tool.

Table 3.1 provides a summarized overall assessment of the previously mentioned security management methodologies based upon the twelve criteria previously described.

All the methodologies and methods presented describe specific implementation steps for the evaluation of the security level of the organizations. Beside ISO 27001 and ISO 27002 that provide generic requirements of risk assessment and do not include specific risk handling aspects, all the other approaches provide well-defined actions and steps for the execution of the risk analysis and risk management processes. Also, most of the methods are meant to be used with qualitative measurements, and this confirms

Table 3.1 Assessment of Security Management Methods and criteria

Methods	Criteria C1	C2	C3	C4	C5
ISO/IEC 27005:2008	General-purpose	Government, agencies, large companies, SME	RA/RM	Quantitative/Qualitative	Based on the business harm
NIST 800-30	General-purpose	Government, agencies, large companies	RA/RM	Qualitative	Based on open damage scenarios
OCTAVE	General-purpose	Government, agencies, large companies, SME	RA/RM	Qualitative	Based on critical assets
CRAMM	General-purpose	Government, agencies, large companies	RA/RM	Qualitative	Based on open damage scenarios
EBIOS	General-purpose	Government, agencies, large companies, SME	RA/RM	Qualitative	Based on security needs
IT-Grundschutz	General-purpose	Government, agencies, large companies, SME	RA/RM	Qualitative	Based on open damage scenarios
MAGERIT	General-purpose	Government, agencies, large companies	RA/RM	Quantitative/Qualitative	Based on open damage scenarios
MEHARI	General-purpose	Government, agencies, medium to large companies	RA/RM	Qualitative	Based on fixed damage scenarios
ISAMM	General-purpose	Government, agencies, large companies, SME	RA/RM	Qualitative	Based on monetary loss

C6	C7	C8	C9	C10	C11	C12
N/A	Low	Low	Standard	Commercial	No	ISO/IEC 27001:2005, ISO/IEC 27002:2005
Type 1	Low	Low	ITC experts	Free	No	ISO/IEC 27001:2005
Type 4	Medium	Low	Standard	Free	Yes/ Commercial	—
Type 1	Low	Low	ITC experts	Commercial	Yes/ Commercial	ISO/IEC 27002:2005
Type 2	Medium	Low	Standard	Free	Yes/Free	ISO/IEC 27001:2005, ISO/IEC 27002:2005, ISO/IEC 27005:2008
Type 5	Low	Low	Standard	Free	Yes/ Commercial	ISO/IEC 27001:2005, ISO/IEC 27002:2005
Type 5	Low	Low	ITC experts	Free	Yes/ Commercial	ISO/IEC 27001:2005, ISO/IEC 27002:2005, ISO/IEC 27005:2008
Type 1	Low	Low	ITC experts	Commercial	Yes/ Commercial/ Free	ISO/IEC 27001:2005, ISO/IEC 27005:2008
Type 3	Low	Low	Standard	N/A	Yes/ Commercial/ Free	ISO/IEC 27002:2005

the fact that most risk assessments today are carried out in a qualitative way, mainly due to lack of reliable quantitative data or to time constraints.

We have identified the following pitfalls of the described methodologies:

Only a part of them are supported by software (in some cases freeware) tools. The general characteristics of these tools are:

- not easy to use without high security expertise;
- monolithic and standalone, thus failing to address advanced requirements of the modern information systems;
- the use of advanced and interactive Web-based graphical user interfaces and the collaboration aspect are two notable requirements that the existing tools and applications do not satisfy. For this reason, most solutions fail to facilitate the distribution and sharing of the information, experience, and expertise within an enterprise and encourage the users to jointly work for the implementation of the phases of the risk analysis and risk management in an effective and smooth manner.

Regarding impact level evaluation [160], the ISO 27005 and ISO 27002 impose that the impact of a security event is assessed in terms of the business harm caused to the organization. In MEHARI, the analysts involved in the risk analysis process measure the impact level based on a "fixed" impact scenario. On the other hand, CRAMM, IT-Grundshutz, NIST SP 800-30, STORM, and MAGERIT give the opportunity to the analysts to specify different impact scenarios (e.g., from catastrophic to marginal) that depict the negative effects of an event (e.g., threat, attack) on the organization. In this context, these approaches adopt the concept of damage scenarios. OCTAVE measures impact based on how "hard" a security event affects a critical asset. In EBIOS, the impact level of an event is measured considering the security needs that the specific event violates. Similarly, in ISAMM the impact is assessed in terms of financial losses the organization has suffered because of an event. The impact level evaluation approach needs to be standardized to achieve uniform evaluations for the EU ports.

They fail to capture the complexity of infrastructure interconnections, cross-sector impacts, dependencies with other systems, or infrastructures and cascading effects within a sector or across sectors. Therefore, various modifications are required to be applied in the security management of the ports' ICT systems since these systems interact with many maritime external entities (interdependencies analysis).

The methodologies try to cover the obligations imposed by the ISO family of standards (ISO/IEC 27001:2005, ISO/IEC 27002:2005, ISO/IEC 27005:2008). However, only EBIOS and MAGERIT achieve full compliance with their rules and procedures. CRAMM, ISAMM, and IT-Grundschutz follow the code of implementation of an ISMS as described by the ISO/IEC 27002:2005, while NIST 800-30, IT-Grundschutz, and MEHARI satisfy the requirements defined by the ISO/IEC 27001:2005. They are very generic, failing to provide targeted technical solutions that address specific sectoral (e.g., maritime) problems and threats, such as interdependent threats rising from associated entities, sector-specific threats (e.g., weather conditions, strikes), and sector-specific legislation (e.g., ISPS in maritime environment). However, these methods and standards provide an insight for the security management of the ports' ICT systems.

All the methods rely on interviews and workshops to aggregate and accumulate the information of the security assessment. However, most of them present limited collaborative capabilities since they do not promote the extensive and efficient collaboration among the involved stakeholders, the effective discussion and exchange of information, ideas, and thoughts, as well as the active involvement of the corporate representatives. This is a major drawback if we want to apply them in the ports' ICT systems since there are many users and the required face-to-face interviews will require time, effort, and resources. Only OCTAVE and EBIOS provide basic collaboration abilities containing specific features that allow the users to participate actively in several parts of the evaluation process.

Regarding risk level evaluation, the following conclusions can be deduced. CRAMM, MEHARI, STORM, and NIST SP 800-30 share the same common view on risk, i.e., they all consider risk as a combination of the likelihood and the impact of a threat to hit a group of assets and the vulnerability level of this group of assets. Similarly, IT-Grundshutz and MAGERIT consider risk as the combination of the likelihood of an incident (i.e., a threat exploiting some vulnerabilities) and the consequences (positive or negative) of this incident happening. On the other hand, Type 2, Type 3, and Type 4 profiles are intrinsically tied to a particular approach to risk analysis, since Type 2 and Type 4 rely on qualitative concepts for defining risk (e.g., critical assets, security needs) and Type 3 relies on the quantitative concepts of probability and average monetary loss. Finally, the methods of the ISO family do not adopt any risk analysis profile. This is because ISO 27005 is a very general guideline to set up a risk management framework, while ISO 27002 and ISO 27001 are not real methods for risk management, but rather compliance standards, reporting a list of controls for good security practices and the requisites that an existing method should have to be standard-compliant, respectively. Another important drawback of the risk analysis approaches is the lack of efficient and advanced computational techniques. They usually rely on primitive methods to evaluate, determine, and mitigate the corporate risks and use ineffective procedures and techniques to analyze and combine the diverse knowledge located in the organizations. The adoption of more advanced approaches (using new techniques, e.g., fuzzy logic, graph theory, group decision-making) that enhance the inherent capabilities of the risk analysis solutions will increase the accuracy of their conclusions.

CIIP METHODOLOGIES

Several standards and methodologies exist for ensuring the critical information infrastructure protection (CIIP). Most of them contain procedures, definitions, and explanations of techniques used to collect and analyze information in CIIP.

Most the CIIP standards and methodologies are from the energy sector and more specifically from the US Department of Energy and North American Reliability Corporation. From its 83 standards a group of specific standards (CIP-002-3 and CIP-003-3) address the protection of critical infrastructures, but only for the electrical energy systems in a very abstract manner. However, since they are very generic, they can provide an insight for the ports. Various national risk assessment methodologies for CIs exist that use various ways in estimating the criticality of an infrastructure even at national level (e.g., in the Netherlands). There is not a standardized way in examining the criticality of an infrastructure or/and addressing cyber threats. A short description of these CIIP methodologies is presented in Table 3.2.

The CIIP methodologies and standards have been evaluated [159] according to the following criteria:

- **availability**: availability of supported applications (under research (R) and/or development (D), or already available for use by the public with commercial purposes (C) or by a limited or restricted group, normally the military (L));
- **CI affected**: The CI sectors that are covered based on [NIPP2009] and Directive 114/08 11 include electricity (1); natural gas (2); oil and pipelines (3); drinking water (4); sewage and wastewater (5); industrial control (6); telecommunications (7); computer networks and information systems (8); railways (9); highways and roads (10); human activities including services and emergency evacuation (11); banking and finance (12); also, the policies and regulations features (13);

Table 3.2 CIIP methodologies short description

	Nature	Description
ATHENA	Software tool	Provides a model for vulnerability analysis of interdependent infrastructure networks
Critical infrastructures interdependencies integrator (CI3)	Software tool	Estimates service's restoration time and cost
Critical infrastructure protection decision support system (CIP/DSS)	Software tool	Provides decision support for CIIP comparing the effectiveness of strategies to reduce the probability of a risk
Critical infrastructure protection modeling and analysis (CIPMA)	Software tool	Evaluates the effects on the operation disruption of CII services within and across sectors
Agent-based simulation model of the US economy (COMM-ASPEN)	Software tool	Provides simulations of the effects of both market decisions and interruptions of telecom infrastructure in the economy based on an agent-based approach
DUTCH NRA	Working methodology	Provides a multicriteria decision-making approach for the risk evaluation considering political and societal issues
Procedimiento informá tico-ló gico para el aná lisis de riesgos (EAR-PILAR)	Software tool	Supports a comprehensive risk analysis method
Electricity market complex adaptive system (EMCAS)	Software tool	Provides an in-depth investigation of the operational and economic impacts on the electrical system, as affected by various external events, based on an agent-simulation approach
Fast analysis infrastructure tool (FAIT)	Software tool	Provides a framework for conducting economic impact assessment across multiple sectors
Financial system infrastructure (FINSIM)	Software tool	It applies to scenarios of crisis affecting the banking payment system, the use of plastic money, the federal funds market, and the interactions between these entities

Continued

Table 3.2 CIIP methodologies short description—cont'd

	Nature	Description
Failure modes and effects analysis (FMEA-FMECA)	Working methodology	Provides a procedural approach for identifying and analyzing possible failures in the design, development and maintenance of a system, based upon the severity or the effect of system failures
FORT-FUTURE	Software tool	Provides a framework that enables decisionmakers to virtually test potential solutions running multiple dynamic simulations
Fault tree analysis (FTA)	Working methodology	Provides a method to failure analysis, identifying the causes leading to the manifestation of a risk within a system
Interoperability (GIS)	Working methodology	Using geographic information systems in emergency coordination and support for decision-making
GORAF	Software tool	Provides a framework for the identification and analysis of the most critical resources within an infrastructure
Hazardous operations (HAZOP)	Working methodology	Support a range of techniques for the identification of potentially hazardous conditions and risks based on assumptions
Inoperability input–output model (IIM)	Software tool	Provides a comprehensive framework based on analytical models to identify and address the risks come from the intra- and interconnectedness of economic sectors
INTEPOINT VU	Software tool	Adopts a decision support model to analyze planning responses to intentional and unintentional events
LUND	Working methodology	Provides a method for the representation of a system of roads or rail interconnected transport infrastructure

Metodología de análisis y gestión de riesgos de los sistemas de información (MARGERIT)	Working methodology	Provides an approach that gives emphasis on the protection of the ICT infrastructure
Methodology for interdependencies assessment (MIA)	Working methodology	Aims to identify and evaluate the interdependencies among critical ICT components
Multilayer infrastructure (MIN)	Software tool	Provides a dynamic game theoretic model to analyze multilayer infrastructure networks
Multinetwork interdependent critical infrastructure programme for analysis of lifelines (MUNICIPAL)	Software tool	Provides a framework for identifying, analyzing, and responding to events that affect the interdependence of civil infrastructure
National agent-based laboratory for economics (N-ABLE)	Software tool	Identifies and analyzes economic factors, feedbacks, and downstream effects of road transport infrastructure and electricity markets
Net-centric effects-based operations model (NEMO)	Software tool	Provides an environment to support decision-making in the area of planning an infrastructure system
Network security risk assessment model (NSRAM)	Software tool	Analyses interconnected multiinfrastructure networks in order to determine the system behavior to various kinds of negative events
Risk maps	Working methodology	Supports a method for identifying and recording of risks in a systematic and effective manner
Transportation routing analysis geographic information system (TRAGIS)	Software tool	Provides a method for the optimization of transportation routes
Urban infrastructure suite (UIS)	Software tool	Provides a simulation-based approach to represent urban infrastructures and populations
Virtual interacting network community (VINCI)	Working methodology	Provides virtualization of the network architecture for critical infrastructures

- **stage**: the functionality provided in each of the stages of risk management programs: identification of assets (a); risk assessment (b); prioritization of actions (c); implementation programs (d); and effectiveness measurement (e).

Based upon these criteria, an overall ranking is described in Table 3.3.

CIIP methodologies mostly address safety threats; over two-thirds of them are implemented in software tools, and the rest are analytical and generic methodologies. About half of the applications have resulted in the development of computer platforms, whether commercial or of restricted use, e.g., corporate, institutional, private, or military. One-quarter of the

Table 3.3 Assessment of suitable CIIP methods

CIIP methodologies	Availability	CI sector	Stage
ATHENA	L	1,2,3,4,5,6,**7**,**8**,**9**,**10**,11,12,13	b
CI3	L	1,2,4,5,6,**7**	c
CIP/DSS	L	1,2,3,4,5,6,**7**,**9**,**10**,11,12,13	a,c,d,e
CIPMA	L	1,2,3,**7**,**8**,12,13	d,e
DUTCH NRA	L	1,3,4,**10**,11,13	a,b,c
EAR/PILAR	C	**8**,11,13	a,b,c,d
EMCAS	C	1,**7**,12	a,b,c
FAIT	L	1,2,5,**9**	a,b
FINSIM	R	**7**,**12**	a,b
FMEA/FMECA	C	6,**7**,11,12	a,b,c
FORT-FUTURE	L	1,2,3,4,5,6,**7**,**9**,**10**,11,12,13	a,b,c,e
FTA	C	6,**7**,11,12	a,b,c
GIS interoperability	R	9,1	c,e
GoRAF	R	1,4,6,**8**,11	b,c,d
HAZOP	C	1,2,3,**7**,11,13	a,b,c
IIM	R	1,4,**7**,**8**,**10**,13	a,c,d
INTEPOINT VU	C	1,**7**,**9**,**10**,11	c
LUND	R	1,**9**,**10**	a,b
MARGERIT V2	C	**8**,11,12,13	a,b,d
MIA	R	**7**,**8**,13	a,b
MIN	R	**10**,**11**	a
MUNICIPAL	R	1,**7**,**8**	a,c
N-ABLE	L	1,**9**,12	a,c,d,e
NEMO	L	1,2,4,**9**,13	c,d,e
NSRAM	R	1,7	c,d,e
Risk maps	R	1,2,3,4,5,6,**7**,**9**,**10**,11,12,13	A
TRAGIS	L	**9**,**1**	A
UIS	L	4,5,**7**,**10**,11	a,b,c,d
VINCI	R	**8**	d

The bold describe the methodologies that can be applied in the transport sector that host information systems.

applications have limited availability and are mostly addressed at military and governmental segments. This could be explained due the leadership taken by some US laboratories, which in turn are sponsored by both the Department of Homeland Security and the Department of Energy. Another quarter of the applications have commercial purposes (computer platform licensing, consulting, etc.). These are extensively used in the energy sector, cybersecurity, and in the definition of emergency response strategies.

Regarding the deployed CI sector, nearly one-quarter of the applications are involved with energy infrastructure (electricity, natural gas, oil, and pipelines). Other infrastructures receiving attention are those related to information technologies and communication and control systems (21%), water (13%), transportation (10%), and banking (8%). About 11% of the methodologies are related to human activities queries and responses checking into CI, which establishes responses to human users' system under emergencies, industrial security, policy recommendations on assets protection, and/or human life protection. Finally, implementation of policies and regulations has attracted special consideration in 12% of the reviewed platforms.

Regarding the risk stage, over two-thirds of the approaches cover the first three steps of a complete risk management process. These steps concern the identification of the corporate assets developing an inventory, the risk assessment process focusing on the evaluation of the security level, and the prioritization of actions establishing priorities for risk assessments, to identify where risk reduction is more compelling and then to determine protective measures that need to be taken. Nearly half of the solutions implement a protection program that includes the deployment of specific protection measures, while only a limited number of the applications apply the stage of measuring effectiveness establishing indicators to provide information on achieving specific security goals.

CYSM RISK ASSESSMENT TOOL AS A BEST PRACTICE

A proposed solution to better address the ports' security is to combine the ISPS with common ICT security management and CIIP standards. However, the well-known security management standards described before will need further modifications to adapt to the port ICT security, so they can address specific sectoral threats, i.e., interdependent threats rising from all entities in the maritime environment, specific threats (e.g., weather conditions, strikes), and maritime legislation (e.g., ISPS).

A targeted security management methodology for the ports' CIs should address the following general requirements:

- *Compatibility with standards*: The methodology needs to comply and implement the ISO 27001, CIIP standards, and the ISPS to address all aspects of the security requirements of the ports' ICT systems that are hosted in the ports CIIs.
- *Collaboration*: Ensure collaboration among all ICT port users (e.g., operators, administrators, collaborators, suppliers, providers).
- *Interdependencies*: All interconnections of the ports' CIIs with all other entities in the maritime ecosystem need to be identified.
- *Broad threat analysis*: Analyze interconnected and interdependent threats.
- *Time and resource economical*: Avoid the plethora of paper-based questionnaires and frustrating interviews imposed by the existing methodologies that are resource (time, personnel) efforts.
- *Easy to implement*: The expert should not need a high level of expertise to apply the methodology.
- *Open*: Avoid security through obscurity.
- *Holistic*: Secure all layers and all assets of the ports' ICT systems.
- *User centric*: Collect and value the opinions and experience of all port users regarding the security of the ports' ICT system.
- *Automation*: Implement the methodology in a user-friendly, open source, collaborative tool.

Any targeted security management methodology needs to be supported by a tool. This tool will be used by the port security team to easily manage their security. Such a tool needs to address the following general requirements:

- *Knowledge codification*: Participants should be able to find relevant risk management knowledge. A codification strategy probably works best for certain types of knowledge that are not expected to change frequently. Participants can then easily retrieve methods and best practices that have proven themselves in the past, and reuse them accordingly.
- *Personalization*: Security management decisions need to reach consensus with all ports' users; knowledge is not always immediately "stable" enough to codify. For such knowledge, a personalization strategy could prove useful to enable participants to find who knows what. Furthermore, personalization techniques should also be valuable to support the discussions and negotiations between ports' users and stakeholders.

- *Collaboration*: Because security management is consensus decision-making, a knowledge-sharing tool should explicitly support collaboration between different users. This property enables the active involvement of all important stakeholders in the decision-making process.
- *Role-specific content views*: Since the security management includes three modes (practicing, criticizing, and reviewing) in which usually different participants' roles are involved, the tool must support specialized views on the available content, such as open issues or approved decisions.
- *Descriptive approach*: Since the security management process is highly creative, the knowledge-sharing tool should not be prescriptive in nature. A more descriptive approach toward the knowledge management would best facilitate the creativity of the participants.
- *Service-oriented*: All steps involved in the security management methodology (boundary specification, asset identification, threat analysis, impact analysis, vulnerability estimation, risk estimation, and mitigation strategy) should be provided as collaborative services.

Cyber/physical security management (CYSM) can serve as a best practice meeting the aforementioned requirements, identifying, classifying, assessing, and mitigating risks associated with port infrastructures and security and safety incidents. CYSM risk management tool has been developed in the EU CIPS Project CYSM under the Horizon2020 program. It has been developed based on several customized and specialized self-management functions that aim to optimize, merge, and enhance the existing approaches identified in the previous section. The section provides an in-depth analysis of the proposed Collaborative CYSM system, presenting the supported functionality and the adopted processes.

The CYSM system is an innovative, scalable Risk Assessment Toolkit that facilitates the ports' security team to efficiently identify, assess, and treat their security and safety incidents involving all port operators and users. The toolkit adopts and implements a bouquet of flexible and configurable self-driven functions and procedures that constitute the conceptual pillars for building a solution that assists ports to improve their current cyber and physical level. In this context, for CYSM to support sound decision-making, it does the following:

- incorporates a conformance approach that checks and defines the compliance of the ports against the requirements, rules, and obligations imposed by a set of security management standards (ISO 27001, ISPS) and the relative security and safety legal and regulatory framework;

- incorporates a collaborative, multiattribute, group decision-making algorithm that collects the diverse security-related knowledge located in the ports and the results (e.g., threats, vulnerabilities metrics, prioritization of countermeasures) produced by the automated and semiautomated risk assessment routines and processes to (1) determine the value of the information assets; (2) identify the applicable threats and vulnerabilities that exist (or could exist); (3) identify the existing controls and their effect on the risk identified; (4) determine the potential consequences; and (5) prioritize the derived risks and ranks them against the risk evaluation criteria set in the context establishment;
- integrates a security policy–growing mechanism that provides a flexible way for creating and updating customized security policies and procedures;
- implements a social, collaborative working environment, which will facilitate and encourage the ports to jointly work and cooperate, by exchanging ideas and information pertaining to security and safety issues and by allowing them to reach targeted solutions in a collaborative and time-effective manner.

The elements are combined in an effective and efficient manner to develop the automated routines and workflows that comprise and construct the meaningful CYSM Security Assessment Services, i.e., Cyber Risk Assessment Services, Physical Risk Assessment Services, and the Security Framework Service. These services are fully customizable depending on the ports' security profile (like the enterprise size, the interdependencies with other IT systems, the services offered, the number of administrators, and the security and safety awareness level), covering various aspects such as complexity, automation, terminology, simplification, and understanding (Fig. 3.1).

CYSM System Components

For the proposed system to meet its objectives, it integrates a set of primary components. From a conceptual perspective, the main components are the following:

- *Community Portal*: This area is accessible by all users of the involved ports in the CYSM community and comprises the following:
  - Community collaboration suite: encapsulates a set of specialized Web2.0 elements (e.g., blogs, forums) suitable for e-collaborate, collecting, and sharing knowledge. These elements enable ports to work together in building open working groups, providing diverse

Figure 3.1 Collaborative cyber/physical security management (CYSM) system.

opinions, thoughts and contributions, and sharing information, experience, and expertise;

- Community e-Library: acts as the knowledge source of all ports' physical- and cyber-related information (e.g., European legal and regulatory framework, security-related standards, specifications, methodologies, and frameworks).

- *Port Private Portal*: This area provides the appropriate functionality that enables the users to assess and improve the security and safety level of their own port infrastructure. This area executes the risk assessment processes and routines integrated in the system and consist of the following modules:
  - Port collaboration suite: encourages and facilitates members of each port to closely cooperate and exchange information and ideas during the risk assessments;
  - Port e-Library: is an inventory of confidential announcements, security and safety policies and procedures, guidelines, etc., that is valuable, private knowledge for the port;
  - Administration module: allows customizing of the risk assessment's parameters (e.g., threats, vulnerabilities, controls);
  - Management module: allows the initiation of a risk assessment;
  - Risk Assessment module: gives the opportunity to the ports to identify and measure their threats, their vulnerabilities, and possible impacts;
  - Security Policy Reporting module: facilitates the formulation of customized security and safety-related policies and procedures;
  - Risk Assessment Results module: allows the review of the risk assessment results and the formulation of a mitigation plan.

These components are provided through customized intuitive and interactive Web interfaces (including interactive screens, online forms, dynamic questionnaires) to represent the scenarios and steps as well as the information and content (e.g., requirements, rules, obligations, and recommendations of the standardization framework and regime) required by the supported risk self-assessment routines and functions, presented in the previous section.

Showcase Scenario

The scope of this section is to describe a use case to illustrate the functionality of the CYSM system. The scenario involves a commercial port, Piraeus Port Authority (PPA), that supports several business operations including the transport and accommodation of people, freight, natural gas, oil, cargos, and manufactured goods. For this reason, PPA manages and operates multiple and dispersed cyber (e.g., computer center) and physical (e.g., facilities for handling all types of cargos) facilities. According to the scenario, the security officer (SeOf) of the PPA utilizes the CYSM system to identify, evaluate, and manage the cyber and physical risks associated with the Cruising Facility of the port and to formulate a mitigation plan. The CYSM system guides and directs the SeOf via dynamic, interactive, and evolutionary interfaces to perform the evaluation process (Fig. 3.2). This process can be divided into the following five phases:

Customization Phase

Initially, the SeOf, authenticates himself (using his credentials) into the CYSM system. Upon approval the SeOf gains access to the Community

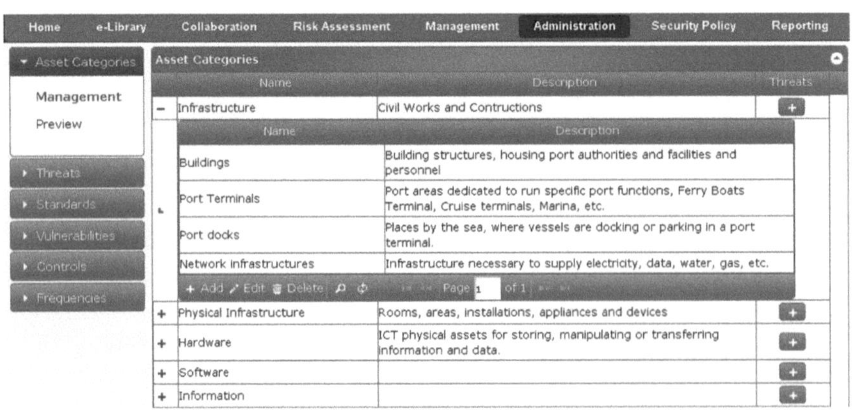

Figure 3.2 CYSM Administration module.

Portal and is directed to the PPA Private Portal where he accesses the Administration module. This module allows the SeOf to set various boundaries and constraints:

- to manage the list of threats that is facing the port;
- to manage the list of vulnerabilities;
- to manage the list of controls that are deployed or can be applied from the organization to mitigate the risks and deal with their defined threats and weaknesses;
- to define the correlation among controls, vulnerabilities, and threats;
- to manage the list of the assets categories and asset subcategories based on which cyber and physical assets will be classified and to define the threats that are applicable to each asset category;
- to manage the standards against which the port's cyber and physical facilities will be assessed;
- to define the correlation between the controls that can be applied from a port and the requirements imposed by the standards;
- to customize the fundamental elements and parameters of the risk assessment procedure (e.g., the scales related to the likelihood of occurrence of the threats, the exploitation level of the vulnerabilities).

The SeOf should update the content taking into consideration the literature, the port's particularities, the adopted technological solutions, the knowledge gained from the daily operation of the port, online repositories available from industry/standardization bodies, national governments, etc.

Risk Assessment Initiation Phase

After the successful customization of the system, the SeOf accesses the Management module where he can initiate a risk assessment (Fig. 3.3). For the definition of a new assessment, the SeOf should specify: (1) the basic information (e.g., name, the start and end date, and a short description); (2) the boundaries of the risk assessment (the physical or ICT port facility that will be assessed); (3) the departments that will be involved and the role and weight of each department to the risk procedure; and finally (4) the standards or the areas of the standards (ISO27001 and ISPS code) against which the defined area will be evaluated.

Evaluation Phase

Having initiated the risk assessment, all the members (participants) of the departments are invited to participate in the risk assessment process, login to the CYSM system using their accounts, and access the Risk Assessment

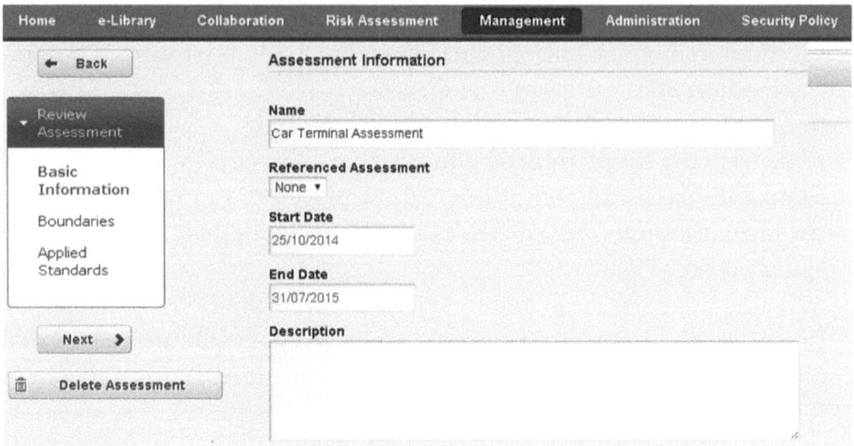

Figure 3.3 CYSM Management module.

Figure 3.4 CYSM Risk Assessment module.

module of the PPA Private Portal (Fig. 3.4). In this module, a list of the available assessments appears, and the participants select the assessment related to the evaluation of the Cruising Facility to complete the following steps:

1. *Assets Identification*: Define the assets that comprise the Cruising Facility and categorize them in the main categories and subcategories defined in the Administration module.
2. *Impact Assessment*: Determine the value of each of them to the organization. They should define what are the consequences (e.g., financial losses, damage to the reputation, legal consequences) of the loss of integrity, confidentiality, and availability of each asset.

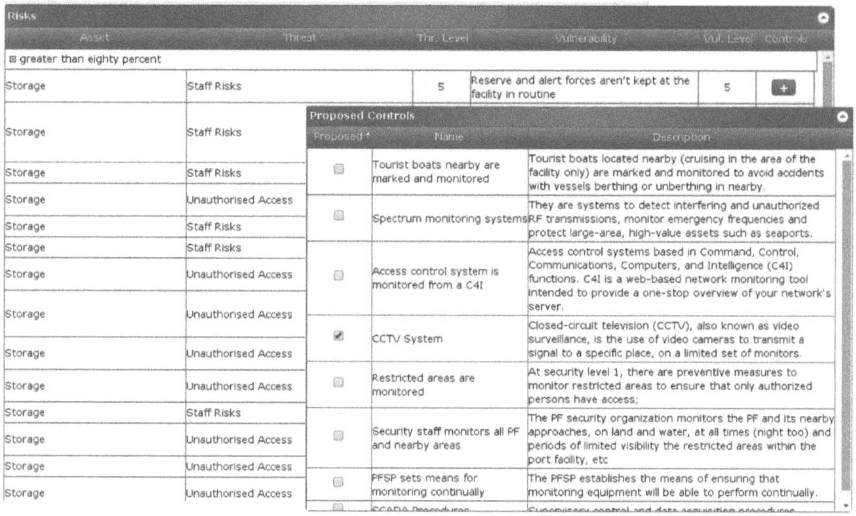

Figure 3.5 CYSM Risk Assessment Results module.

3. *Control Identification*: Define the controls applied to each asset.
4. *Threat Assessment*: Estimate the likelihood of occurrence of a predefined list of threats to each asset.

Risk Assessment Results Calculation and Review Phase

Once all the participants completed the evaluation phase, the SeOf accesses the Risk Assessment Results module (Fig. 3.5) to produce and review the results of the risk assessment. More specifically, the SeOf selects the assessment that he is interested in and forces the system to calculate the potential risks associated with the Cruising Facility, taking into consideration the answers of the participant who completed the evaluation. Now, the SeOf can review the produced results based on those he can select and prioritize the countermeasures that should be adopted in the PPA to handle and mitigate the identified risks. In this way, the SeOf can formulate an effective and efficient risk mitigation plan.

Security Policy Reporting Phase

Finally, the SeOf accesses the Security Policy Reporting module (Fig. 3.6) to formulate the security and safety policies required by the existing regulatory regime (ISO27001 and ISPS code). These policies can be exported in various formats (e.g., PDF, TXT, JPG).

CYSM system is a security management revolutionary consultation environment that is oriented to the special requirements of ports and is in

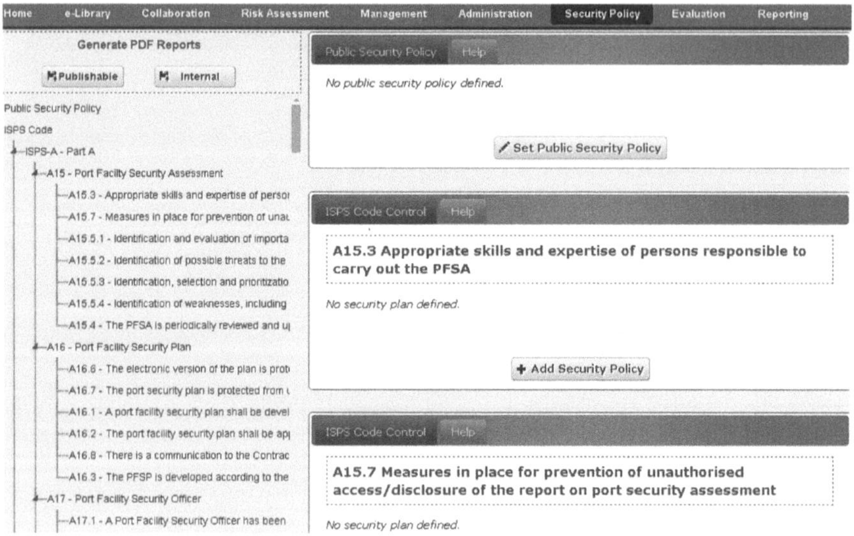

Figure 3.6 CYSM Security Policy Reporting module.

accordance to the basic principles and the business goals of existing risk assessment standards and methodologies. The nature of the system is associated with a high degree of innovation since it implements new upgrading security and safety self-management functions and processes for the evaluation and mitigation of the risks and threats associated to the ports' infrastructure.

CYSM Architecture

The CYSM Risk Assessment Toolkit adopts and implements the functions and procedures presented in the previous sections. For the CYSM Risk Assessment Toolkit to meet its objectives, it incorporates and merges a set of integrated and interconnected subsystems (Fig. 3.7). From a conceptual perspective, the required subsystems are the following:

The **Web Interactive component** provides an intuitive, interactive, and graphic way (e.g., dynamic forms) to represent the information and content (e.g., requirements, rules, obligations, recommendations, and advices of the legal, regulatory, and standardization framework and regime as well as security and safety content required for automating technical control compliance, vulnerability checking, and security measurement activities) of the risk assessment services. It is designed to be a single environment where all the applications a user needs can run, and these are integrated together in a consistent and systematic way.

Figure 3.7 CYSM Risk Assessment Toolkit architecture.

If an application lives outside of the main CYSM portal, the portal should be able to consume some resource of the application (such as an RSS feed or a subset of functionality in a "dashboard" application), so the end user will be able to see everything he/she interacts with at a glance.

To achieve this, all the application functionality within **CYSM Web Interactive component** is in fragments of the page, called portlets. Portlets are Web applications that run in a portion of a Web page. Within the proposed platform, the portlet container is the most important architectural bone. The container's job is to aggregate the set of portlets that are to appear on any page and display them properly to the user. In this way, one or many applications reside on a page, and the user can (at the administrator's discretion) arrange them in the way that works best for the user.

Portlet applications, like servlet applications, have become a Java standard, which various portal server vendors have implemented. The Java standard defines the portlet specification. A JSR-168 or JSR-286 standard portlet is deployable on any portlet container, which supports those standards. Portlets are placed on the page in a certain order by the end user and are served up dynamically by the portal server.

Portal applications come generally in two flavors:

1. multiple portlets written to provide small amounts of functionality and aggregated by the portal server into a larger application;
2. whole applications written to reside in only one or a few portlet windows.

Portlets are not difficult to build, and Java standard portlets can be written by any Java developer with experience in writing Web applications.

CYSM platform provides a Plugins Software Development Kit that makes the creating of portlet projects easy.

Platform administrators can use the Dockbar, which is the primary tool logged in that users have for navigating the portal and accessing administrative functions from anywhere on the Web site. Depending on the logged-in user's roles and what section of the CYSM portal the user is viewing, all or only some of the options are available in the Dockbar.

Moreover, administrators can use the Manage menu in which they are able to access various settings for the current page and any of its subpages. The items available are Page, Page Layout Sitemap, and Settings. With Page Layout a user can choose the layout template to use for the current page. The other settings are the same as their counterparts in the Control Panel, which is also an important option for configuring appropriately the portal and its functionality.

With the usage of the Toggle Edit Controls a user can turn on and off the edit controls of the portlet windows. This is helpful for administrators who want to look at a page they are working on and see it the way a regular user would.

Additionally, in the Dockbar a user can be informed of all the places in the portal to which he has access. CYSM platform allows for various configurations of pages for end users: configure it so that some or all users have their own pages, public and private (or both), upon which they can place the portlets they need to use. The administrator account by default has its own pages.

One of the most important tools CYSM platform offers for managing the portal is the Control Panel. The Control Panel is composed of administrative portlets that a user can use to manage various aspects of the portal.

The **Semantic Modeling component**, on the other hand, integrates a collection of semantic structures (notably ontologies/taxonomies) modeling:

- the security and safety posture of the ports;
- the employees of ports based on their cognitive state and behavior regarding their role and responsibilities in their enterprise, existing background knowledge about security and safety, and level of their interaction with the port infrastructure;
- all the cyber and physical aspects (e.g., security and safety-related legal, regulatory and standardization framework and tools, etc.);
- their semantic relationships providing a modularization of knowledge.

This module is responsible for the content categories that are defined by someone with administrative access to the content. They are hierarchical, tree-like structures that users can use to find content. Categories are different from tags in that they are never created by end users. Instead, categories define how the content is organized from the point of view of the owner of the content (e.g., ports). A good example of categories might be the organization of a port: it shows the hierarchical structure and organization for all the departments within that port.

The **Execution Engine component** incorporates an automated workflow tool that executes all the underlying complex processes and routines defined by the proposed CYSM Risk Management Methodology (CYSM-RM) of the services, providing a high degree of automation and transparency. Also, this component undertakes the responsibility using elements of the Web Interactive component (e.g., online forms) to guide and direct the users to perform the required activities and actions.

CYSM workflows are essentially a predetermined sequence of connected steps. In a CYSM system, each workflow is designed to manage the creation, modification, and publication of Web content. An administrator can set up a workflow so that content cannot be published without going through an approval process designed. In this way, content goes up on the site only after it has been reviewed and approved.

The CYSM system's workflow engine allows a privileged user to define any number of simple to complex business processes/workflows, deploy them, and manage them through a portal interface (Web Interactive Tier). Those processes have knowledge of users, groups, and roles without writing a single line of code; it only requires the creation of a single XML document. This document is executed by the end users on the portal. Administrators can create as many different workflow definitions as they need to manage the work done on their portal.

The key parts of the workflow definition are the asset, states, transitions, and tasks. The asset is whatever piece of content is being reviewed and approved in the workflow. States represent stages of the workflow, for example, created, rejected, or approved. Transitions occur between states, and indicate what the next state should be. Tasks are steps in the workflow that require user action.

The **Interaction component** embodies mechanisms and interfaces that implement a set of standards and languages to encapsulate information from other services and automated tools in an automated and transparent way. This component integrates all the characteristics of an enterprise service bus, which is a software architecture construct that lives between the (business) applications, enabling communication among them. Ideally, it replaces all direct contact with the applications on the component itself, so all communication takes place via it.

The Interaction component provides its fundamental services via an event-driven and standards-based messaging engine (the bus). Through its utilization, integration architects can exploit the value of messaging without writing code. Developers can use technologies found in a category of middleware infrastructure products, usually based on recognized standards.

With this component, a potential author such as the CYSM technical team can filter, transform, route, and manipulate SOAP, binary, plain XML, and text messages that pass through his business systems by HTTP, HTTPS, JMS, mail, etc.

Finally, the **Advanced Security Intelligence Engine** delivers analysis of all activity observed within the ports' environment in an effective and efficient manner. The component incorporates the proposed CYSM Risk Management Methodology (CYSM-RM) and a set of technologies for enumerating, describing, measuring/quantifying, and encapsulating data (e.g., findings, threats and vulnerabilities metrics, and prioritization of countermeasures). With a practical combination of flexibility, usability, and comprehensive data analysis, the proposed engine delivers visibility to risks, threats, and critical operations issues. Also, it provides a full lifecycle of security and safety management by incorporating the following components:

- The Asset Identification component encompasses the mechanisms required to collect and gather the critical information and physical assets of the evaluated ports' facilities. These aspects are not confined only to technical issues, but they are also concerned with the business and physical processes in which the systems are embedded.
- The Evaluation component integrates the appropriate means to assess the security and safety of ports' operational environment. This component conducts an analysis to pinpoint threats and vulnerabilities and assigns a rank to each based on the risk potential verses consequences. Finally, it can make recommendations on how best to minimize against these consequences.

The Asset Identification component plays an important role in all integrated CYSM services and especially in the Risk Assessment service. It is required from each port to quickly correlate different sets of information about its assets. The CYSM platform provides the necessary constructs to uniquely identify assets based on known identifiers and/or known information about the assets. Within the CYSM platform for the asset identification, different categories of assets have been already defined and properly integrated. These are as follows:

- infrastructures (e.g., buildings, terminals)
- physical assets (lockers, cabinets, etc.)
- hardware assets (e.g., servers, routers, switches)
- software assets (operating systems, applications, etc.)
- data assets (e.g., electronic data, paper-based/printed data)

The Evaluation component, on the other hand, refers to the risk treatment plan, which is part of the risk management plan for each port. It outlines how risks will be managed whether they are low, high, or acceptable risks according to the management decision of each port.

The CYSM system serves as a best practice example of a targeted (physical and cyber) risk management tool for ports' operators. It adopts a simplified and optimized approach as a response to the transitional time-consuming risk assessment procedures. The proposed approach is represented through several customized and specialized self-risk assessment processes that are modeled and implemented in the system in a graphic approach using visualization tools, automated routines, processes, and structured content that encapsulate several simplifications and incorporate automated steps. CYSM offers open source "easy to use" tools enabling the security and safety management in intuitive and graphic way. In a nutshell, the basic design principles of all the related functions developed within CYSM include these:

- *Easy to use for nonexperts.* Simplification and automation of the supported risk assessment procedures and activities makes it possible for the ports' personnel to conduct self-assessments.
- *Self-driven.* Deployment of the procedures without the need of external resources. The personnel of the ports are guided and directed through automated workflows and routines and online intuitive and interactive graphic representations (e.g., dynamic questionnaires) to use the provided functionality.
- *Collaborative.* The risk assessment process is treated as a participatory challenge among various user groups with a different vision angle for each one and different standards, respectively. In this context, CYSM calculates the actual risk factor based on the opinion of all involved users and not only to persons that are involved with security and safety,

considering their experience and expertise and the different points of observations. CYSM raises the security awareness and consciousness of all port operators and users.

The CYSM system has been tested and evaluated by several commercial ports (including Port of Piraeus, Valencia Port Authority, and Port of Mykonos) during the CYSM project (in the pilot operation phase). During the evaluation operation, various ports' users (e.g., port security officers, members of ports' security teams, ports administrators, and internal users interacting with ports' ICT systems) have been engaged in risk identification, assessment, and mitigation based on the online services of the CYSM system.

Finally, CYSM promotes a holistic approach to the security and safety management for ports. This approach is based on flexible semantic infrastructure and tagging mechanisms for security policies and other relative security and safety data. The provision of such structured knowledge enables ports to handle their cyber and physical issues in an effective and unified manner and facilitates their cooperation, encouraging working together on issues of mutual concern.

CYSM is not a product but a research initiative that can serve as a best practice in the development of new ICT security management tools that are compliant with the ISO 27001, CIIP, and ISPS, helping the port operators to better manage their physical and cyber risks within their CIIs.

CHAPTER 4

Maritime Supply Chain Risk Assessment (at Entity Level)

Most of the risk assessment methodologies we saw in the previous chapters focus on the identification of threats, vulnerabilities, and the related impact and ultimately on the evaluation of the underlying risks within the boundaries of an organization or a critical information infrastructure (CII).

However, most risk assessment methodologies are organization-oriented and do not consider the cascading effects occurring from cross-sectoral and/or cross-border dependencies of involved entities. Consequently, they tend to focus on organization-wide risks and they fail to capture the security needs of complex supply chain services (SCSs), such as the ones described in Chapter 2, and their interdependent entities.

Obviously, there is a need for extending and validating existing risk assessment frameworks in terms of their ability to deal with SCS risk assessment, including the dependencies associated with the provisioning of the SCS involving various entities. A starting point for modeling cascading effects lies in the understanding of the interdependencies of the various infrastructures, which include physical, cyber, geographic, logical, and other dependencies.

Risk management international standards range from general considerations and guidelines for risk management processes to specific guidelines for the IT sector and to critical infrastructure (CI) dependency analysis risk assessment methodologies all the way to sector-specific frameworks, such as in the maritime sector. Most of these standards specify framework conditions for the risk management process but do not provide specific methodologies targeted to SCS risk assessment. In principle, choosing the right method and tool for risk assessment proves to be complicated. Since SCSs originate from a business context, SCS risk management methods are usually quantitative and are based on monetary costs or potential impact. In practice, the selection of a specific risk assessment tool is based on practical considerations, and it depends on how a specific risk assessment methodology can be mapped to the extended risk assessment needs of supply chains.

ISO 28001 is the security management standard specifically developed for supply chain (SC) security, due to an increased demand from

Port Cybersecurity
ISBN 978-0-12-811818-4
http://dx.doi.org/10.1016/B978-0-12-811818-4.00004-6

the transportation and logistics industry. It specifies the requirements for an SC security management system, including aspects related to financing, manufacturing, information management, and the facilities for packing, storing, and transferring goods between modes of transport and locations.

ISO 28001 is applicable to all sizes of organizations in manufacturing, service, storage, or transportation at any stage of the SC. The standard defines the need for certification by an accredited third party and/or a self-declaration of conformance with the standard. ISO 28001 describes a generic methodology at the entity level (it does not go down to interacting cyber assets) and provides a generic guideline for organizations seeking to implement or refine a specific methodology. Finally, NIST IR 7622 aims to provide a notional set of SC assurance methods and practices that offer a SC security management for federal information systems. Although the need for a SC risk management methodology is recognized, the document itself does not define such a methodology.

SUPPLY CHAIN GRAPH MODELS

The definition of a risk assessment methodology targeted to the assessment of SCSs requires several steps: (1) model the supply chain; (2) collect the required input from the involved entities; (3) define threats that are relevant to supply chains; (4) define the appropriate security controls; and (5) define the appropriate scales for all the risk assessment parameters.

We will use the general notation and concepts according to the ISO 28001 and ISO 27001 standards.

Supply Chain Service Graph

SC risk assessment relies on SC-directed dependency graphs, used to visualize the dependencies within an SC (see Fig. 4.1 for an example of an SC graph):

Order is defined with respect to the node $\times 0$ (either as a source or as a destination node).

A supply chain directed graph is a directed graph SCG = ($\mathbb{N}$, $\mathcal{E}$), where $\mathbb{N}$ is the set nodes, and $\mathcal{E}$ is the set of edges.

A node $x_i \in \mathbb{N}$ represents a business partner participating in an SC required for the provision of a service (SC service).

A directed edge $x_i \rightarrow x_j \in \mathcal{E}$ represents a business process provided by x_i to x_j to support the provisioning of an SC service.

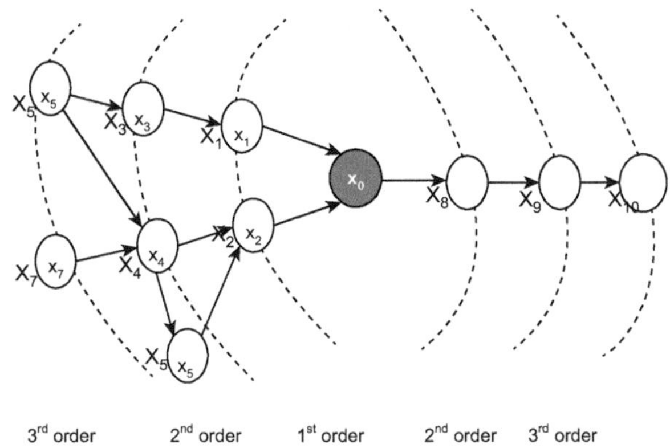

Figure 4.1 An example of an SC-directed graph.

The weight of x_i (w_i) is the importance of x_i for the provisioning of the SC service. It represents the business impact of a business partner (node) for the provisioning of the SC service. The following scale is used:

$w(x_i) = 1$ (High) if its absence/disruption of its operations results in severe delay/disruption of the provisioning of the SCS.

$w(x_i) = 0.5$ (Medium) if its absence/disruption of its operations results in an important delay/disruption of the provisioning of the SCS.

$w(x_i) = 0.25$ (Low) if its absence/disruption of its operations results in limited delay/disruption of the provisioning of the SCS.

Type of dependency: Four types of dependencies are defined between business partners x_i and x_j.

$D(x_i, x_j) = 1$ if access to cyber-systems and e-data: The access could be to a database, to operational systems, networks, etc.

$D(x_i, x_j) = 2$ if interaction with cyber-systems and e-data: The interaction could be by sharing information, the offer of common services, etc.

$D(x_i, x_j) = 3$ if access to physical facilities: These facilities are buildings, terminals, storage rooms, etc.

$D(x_i, x_j) = 4$ if usage of physical facilities: The use could be for warehousing, for hosting an installation, etc.

Order of dependency, $order(x_i, x_j)$: The order of a business partner x_j in relation to another business partner x_i is defined as the number of steps required for x_i to reach x_j using the smallest length (shortest) path: $order(x_i, x_j) = \min(\{ | path(x_i, x_j) | \})$.

For example, in Fig. 4.1, we have $order(x_0, x_8) = 1$ and $order(x_5, x_0) = 2$.

Security Declaration Statements

ISO 28001 imposes the requirement that each business partner within an SCS maintain a *Security Declaration Statement* (SDS), a documented commitment that defines the security measures implemented by each business partner, related with the security assurance of the examined SCS. The SDS as defined in ISO28001 can be extended in the following ways:

1. It assigns each statement with a security requirement a unique ID. Then each security requirement will be mapped to the corresponding security vulnerability and security control, following the same ID. For example, the security requirement $R_{1.1}$ will correspond to a security vulnerability $V_{1.1}$ and security control $C_{1.1}$.
2. It allows business partners to declare partial implementation of controls.

Part of an SDS is shown in Table 4.1.

Security Requirements, Vulnerabilities, and Controls

Based on the security requirements for SCS security, as defined in the Declaration Statement described previously, a complete list of security vulnerabilities and their corresponding security controls are provided. A list of security vulnerabilities and security controls that are related with the SCS under assessment is defined. The list of security vulnerabilities and controls, in accordance with ISO 28001, can be categorized in seven security areas: (1) management of SC security, (2) security plan, (3) asset security, (4) personnel security, (5) information security, (6) goods and conveyance security, and (7) closed cargo transport units security. Table 4.2 presents part of the list associated with the area of ICT security.

Table 4.1 Part of a security declaration statement

Implementation level factor (security controls)	Yes	No	Partially	Comments
(1) Management of supply chain security				
$C_{1.1}$ Does the organization have a management system that addresses supply chain security?				
$C_{1.2}$ Does the organization have a person designated as responsible for supply chain security?				

Table 4.2 Security vulnerabilities and the corresponding security controls

Security area: (5) information security			
Security vulnerability		Security control	
$V_{5.1}$	Lack of procedures to ensure that all information used for cargo processing, both electronic and manual, is legible, timely, accurate, and protected against alteration, loss, or introduction of erroneous data	$C_{5.1}$	Implementation of procedures to ensure that all information used for cargo processing, both electronic and manual, is legible, timely, accurate, and protected against alteration, loss, or introduction of erroneous data
$V_{5.2}$	Lack of appropriate shipping documentation reconciling the cargo, when an organization is shipping or receiving cargo	$C_{5.2}$	Implementation of procedures to ensure that an organization shipping or receiving cargo, reconciles the cargo with the appropriate shipping documentation
$V_{5.3}$	Lack of procedures to ensure that cargo information received from business partners is reported accurately and in a timely manner	$C_{5.3}$	Implementation of procedures to ensure that cargo information received from business partners is reported accurately and in a timely manner
$V_{5.4}$	Lack of a data backup process and storage system(s) for relevant data	$C_{5.4}$	Implementation of a data backup process and storage system(s) for relevant data
$V_{5.5}$	Users are not required to have unique identifiers for their personal and sole use	$C_{5.5}$	Implementation of procedures to ensure that all users have a unique identifier (user ID) for their personal and sole use, which can be used to trace their activities
$V_{5.6}$	Lack of an effective password management system to authenticate users and to ensure at least annual password change	$C_{5.6}$	Implementation of an effective password management system to authenticate users and to ensure that the users change their passwords at least annually
$V_{5.7}$	Lack of protection against unauthorized access to and misuse of information	$C_{5.7}$	Implementation of procedures and/or technical controls to ensure protection against unauthorized access to and misuse of information

Supply Chain Threat Scenarios

A threat categorization provides a systematic definition of threat categories, so (1) individual threat scenarios can be systematically identified and categorized for each SCS, in a structured manner, and (2) threat scenarios can be effectively mapped to the appropriate security controls and evaluated for their vulnerability in each business partner participating in the SCS. Medusa utilizes the threat categories defined in ISO 28001 and extends the examined threat scenarios as described in Table 4.3. Note that this categorization is not distinctive, and several threat scenarios may partially belong to more than one category.

Security Controls to Threat Scenarios

A static mapping can be used to correlate the security threat scenarios to their corresponding security vulnerabilities and required security controls. After the security vulnerabilities and the examined threat scenarios have been defined, the mapping of threat scenarios to security vulnerabilities and controls are presented in Table 4.4.

Risk-Related Scales

We use a five-level Likert scale to describe the threat level for an examined threat scenario. When selecting the most appropriate threat level, a business partner will consider the expected probability of occurrence of the examined threat scenario within its assets and premises (Table 4.5).

The likelihood of occurrence of a threat scenario takes as input a threat level and a vulnerability level and outputs the likelihood of occurrence for the threat scenario within a business partner. The likelihood scale can be developed as follows in Table 4.6.

We also need to set a consequence scale. For example, we may define five impact (consequence) classes, ranging from "Minimal" to "Very Significant," corresponding to cardinal numbers 1–5. Each impact class is then assigned to a value range (in a [1–100] scale) and to a default value. In the proposed classification table, the societal impact is aligned with the disruption of the SCSs (Table 4.7).

The risk level of a threat scenario takes as input a likelihood level and a consequence level and outputs the risk of an examined threat scenario for the business partner in question. The risk scale defined also needs to be defined; an example is shown in Table 4.8.

We need to emphasize that the ISO 8001 standard does not impose a standard formulation of the tables (impact, likelihood, risk, etc.).

Table 4.3 Assigning threat scenarios to threat categories

Threat category	Defined threat scenarios: Threat scenario	Defined threat scenarios: Application
TC-1. Infrastructural threats	$TS_{1.1}$: Destroy a major/critical SC infrastructure	• Warehouses of the stored cargo have been bombed • Fences/exterior walls of the warehouses have been destroyed or bypassed • Buildings hosting a data center used in the SCS have been destroyed due to a deliberate action or a physical threat
	$TS_{1.2}$: Suspected or confirmed unauthorized access to SC infrastructures	• Unauthorized access to storage buildings, etc. • CCTV/DVS cameras do not operate well (due to a physical attack or lack of maintenance)
TC-2. Information and ICT threats	$TS_{2.1}$: Information tampering[a]	• Locally or remotely gaining access to the SC's information/documentation systems for the purpose of disrupting operations or facilitating illegal activities
	$TS_{2.2}$: Information loss	• Cargo/shipping/billing/documentation/information is destroyed due to a deliberate attack (e.g., sabotage) or physical attack (e.g., fire)
	$TS_{2.3}$: Communication interruption or loss	• Optical fibers have been smuggled • Network connection has been disrupted
	$TS_{2.4}$: Software/system abuse	• A critical software for the SC has been hacked • Backdoors identified in an SC-critical system
TC-3. Threats related with personnel security and safety	$TS_{3.1}$: People under attack	• SC key personnel have been taken hostages • Threat against the life of people (business partner's personnel, people using the SC, etc.) • Take hostages/kill people
	$TS_{3.2}$: Misuse/abuse of SC procedures	• The employees are not trained in the SC procedures • An employee misuses his/her security credentials • Absence of key personnel of a business partners (e.g., due to a strike)

Continued

Table 4.3 Assigning threat scenarios to threat categories—cont'd

Threat category	Defined threat scenarios Threat scenario	Application
TC-4. Threats related with goods and conveyance security	$TS_{4.1}$: Intrude and/or take control of an asset (including conveyances) within the SC[a] $TS_{4.2}$: Use the SC as a means of smuggling[a] $TS_{4.3}$: Cargo integrity[a] $TS_{4.4}$: Unauthorized use[a] $TS_{4.5}$: Goods and conveyance misuse	• Damage/destroy an asset (including conveyances) • Damage/destroy outside target using the asset or goods • Cause civil or economic disturbance • Illegal weapons into or out of the country/economy • Terrorist into or out of the country/economy • Tampering, sabotage, and/or theft for the purpose of terrorism • Conducting operations in the international SC to facilitate a terrorist incident including using the mode of transportation as a weapon • The cargo received or delivered by/to a wrong person due to the lack of appropriate authentication procedures • The cargo transport-related procedures have been alternated, and the activities have been misused • Closed cargo has been unsealed illegally • Closed cargo contains wrong material • An employee or business partner in the SC is stealing SC goods • Unauthorized access to all cargo and conveyance storage areas

[a]As defined in ISO 28001.

Table 4.4 Assigning security vulnerabilities and security controls to threat categories and their related threat scenarios

Threat category	Threat scenarios	Related vulnerabilities	Required security controls
TC-1: Infrastructural threats	$TS_{1.1}$: Destroy a major/critical SC infrastructure	$V_{1.1}, V_{1.2}$ $V_{2.1}, V_{2.2}, V_{2.3}$	$C_{1.1}, C_{1.2}$ $C_{2.1}, C_{2.2}, C_{2.3}$
	$TS_{1.2}$: Suspected or confirmed unauthorized access to SC infrastructures	$V_{3.1}, V_{3.2}, V_{3.3}, V_{3.4}$	$C_{3.1}, C_{3.2}, C_{3.3}, C_{3.4}$
TC-2: Information and ICT threats	$TS_{2.1}$: Information tampering	$V_{1.1}, V_{1.2}$	$C_{1.1}, C_{1.2}$
	$TS_{2.2}$: Information loss	$V_{2.1}, V_{2.2}, V_{2.3}$	$C_{2.1}, C_{2.2}, C_{2.3}$
	$TS_{2.3}$: Communication interruption or loss	$V_{5.1}, V_{5.2}, V_{5.3}, V_{5.4}, V_{5.5}, V_{5.6}, V_{5.7}$	$C_{5.1}, C_{5.2}, C_{5.3}, C_{5.4}, C_{5.5}, C_{5.6}, C_{5.7}$
	$TS_{2.4}$: Software/system abuse		
TC-3: Personnel security and safety	$TS_{3.1}$: People under attack	$V_{1.1}, V_{1.2}$, $V_{2.1}, V_{2.2}, V_{2.3}$	$C_{1.1}, C_{1.2}$, $C_{2.1}, C_{2.2}, C_{2.3}$
	$TS_{3.2}$: Misuse/abuse of SC procedures	$V_{4.1}, V_{4.2}, V_{4.3}, V_{4.4}$	$C_{4.1}, C_{4.2}, C_{4.3}, C_{4.4}$
TC-4: Goods and conveyance security	$TS_{4.1}$: Intrude and/or take control of an asset (including conveyances) within the SC	$V_{1.1}, V_{1.2}$	$C_{1.1}, C_{1.2}$
	$TS_{4.2}$: Use the SC as a means of smuggling	$V_{2.1}, V_{2.2}, V_{2.3}$	$C_{2.1}, C_{2.2}, C_{2.3}$
	$TS_{4.3}$: Cargo integrity	$V_{3.1}, V_{3.2}, V_{3.3}, V_{3.4}, V_{3.5}$	$C_{3.1}, C_{3.2}, C_{3.3}, C_{3.4}, C_{3.5}$
	$TS_{4.4}$: Unauthorized use	$V_{6.1}, V_{6.2}, V_{6.3}, V_{6.4}, V_{6.5}$	$C_{6.1}, C_{6.2}, C_{6.3}, C_{6.4}, C_{6.5}$
	$TS_{4.5}$: Goods and conveyance misuse	$V_{7.1}, V_{7.2}, V_{7.3}, V_{7.4}$	$C_{7.1}, C_{7.2}, C_{7.3}, C_{7.4}$
TC-5: Other		$V_{1.1}, V_{1.2}$ $V_{2.1}, V_{2.2}, V_{2.3}$ ...	$C_{1.1}, C_{1.2}$ $C_{2.1}, C_{2.2}, C_{2.3}$ ...

Table 4.5 Threat scale

Threat scale			Description of each threat level	
Threat level	**Value range (%)**	**Default value (%)**	**Probability of occurrence**	**History of incidents**
Very high (5)	(80–100]	100	This threat scenario is expected to occur within the assets of the business partner with very high probability (more than 80% probability)	This threat scenario was realized more than once in the last year (12 month period)
High (4)	(60–80]	80	This threat scenario is expected to occur within the assets of the business partner with high probability (61%–80% probability)	This threat scenario was realized once in the last 1 year (12 month period)
Medium (3)	(40–60]	60	This threat scenario is expected to occur within the assets of the business partner with medium probability (41%–60% probability)	More than one incident of this threat was realized in the last 2 years
Low (2)	(20–40]	40	This threat scenario is expected to occur within the assets of the business partner with low probability (21%–40% probability)	At most one incident of this threat was realized in the last 2 years
Very low (1)	[1–20]	20	This threat scenario is expected to occur within the assets of the business partner with very low probability (at most 20% probability)	At most one incident of this threat was realized in the last 3 years

Table 4.6 A likelihood scale

Likelihood scale			Description of likelihood levels
Likelihood level	**Value range**	**Default value (%)**	**Likelihood of occurrence**
Very high likelihood (5)	(80–100]	100	The examined threat scenario is very likely to occur in the business partner
High likelihood (4)	(60–80]	80	The examined threat scenario is likely to occur in the business partner
Medium likelihood (3)	(40–60}	60	The examined threat scenario has medium likelihood to occur in the business partner
Low likelihood (2)	(20–40]	40	The examined threat scenario is unlikely to occur in the business partner
Very low (1)	[1–20]	20	The examined threat scenario is very unlikely to occur in the business partner

Table 4.7 A consequence scale

Consequence (impact) scale			Scales for each impact category			
Consequence level	**Value range**	**Default value**	**Human impact (quantitative scale)**	**Economic impact (quantitative scale)**	**Environmental impact (qualitative scale)**	**Social impact (qualitative scale)**
Very significant (5)	(80–100]	100	Massive loss of life (>2 casualties, >10 severely injured)	Major damage to a critical asset and/or infrastructure preventing further operations (>100 million)	Major destruction of the ecosystem over a large area (>100 million)	Severe damages to the reputation/public confidence –disruption of the SC service for a very long period (>1 week)
Significant (4)	(60–80]	80	Loss of life (at least 1 casualty)	Significant damage [50–100 million)	Significant destruction [50–100 million)	Significant disruption [2 days–1 week)
Average (3)	(40–60]	60	Severe injuries (>5 severely injured but no loss of life)	Important damage [10–50 million)	Important destruction [10–50 million)	Medium disruption [12 h –2 days)
Minor (2)	(20–40]	40	Some injuries (>5 injured but not severely)	Considerable damage [1–10 million)	Considerable destruction [1–10 million)	Short to medium time disruption [3–12 h)
Minimal (1)	(1–20]	20	Minimal injuries (≤5 people injured but not severely)	Some damage [0–1 million)	Some destruction [0–1 million)	Short time disruption (<3 h)

Table 4.8 A risk scale

Risk scale			Description of risk levels
Risk class	**Value range**	**Default value**	**Risk of a threat scenario being realized in a target business partner**
Very high risk (5)	(80–100]	100	The examined threat scenario has very high risk for the business partner
High risk (4)	(60–80]	80	The examined threat scenario has high risk for the business partner
Medium risk (3)	(40–60]	60	The examined threat scenario has medium risk for the business partner
Low risk (2)	(20–40]	40	The examined threat scenario has low risk for the business partner
Very low risk (1)	[1–20]	20	The examined threat scenario has very low risk for the business partner

MEDUSA: A MARITIME SCS RISK ASSESSMENT METHODOLOGY

Medusa was an European Commission project under the CIPS program that developed a methodology and a security management tool with the following characteristics:

- *Holistic view*: Medusa aims to provide a holistic SC view to identify global SC threats, such as the *cascading threats* within the SC. These may not always be easy to identify from an organization-centric perspective.
- *Collaborative*: Medusa aims to promote collaboration between business partners.
- *Compliance with standards*: Medusa will be compliant with a range of existing standards, such as [11–13,26], as a means of increasing its adoption and longevity.
- *Business-centric*: The importance of the business partners in the provision of the SCSs is considered (providing more targeted risk assessment results).
- *Implementable*: The Medusa methodology adopts a sequential step-by-step approach with clear inputs and outputs, so it can be easily implemented in an ICT tool.

- *Sector-independent*: Medusa methodological steps could be applied in other sectors (e.g., business, health, transport) where SCSs are offered.
- *Auditable*: The results produced may be easily compared with other risk assessment methodological approaches since it uses standardized notations.

Initially, the business partners (or the risk assessor that initiates the risk assessment in consensus with the business partners) shall define the scope of the SCS under examination and define the goal, the scope, and the outcome of the SCS. The risk assessor may create use-cases to clarify the business processes of the SCS and to clarify its business role and functions within it. The outcome of this step is a *Statement of Application* that defines the SCS under examination and its boundaries.

Step 1: Analysis of the SCS

In this step, the SCS is analyzed and modeled using graphs.

Step 1.1: Creating a SCS Graph Model

Business process modeling of the SCS is based on supply chain graphs (SCG). The business partners will provide input related with their business role and dependencies within the SCS. By using this input, the assessor will design the SCG corresponding to the SCS.

Step 1.2: Filling & Signing Business Partners Declaration Statements

In this step, all business partners will provide their completed and signed SDSs. The business partners will provide their SDSs that will be used by the Medusa methodology, to assess the vulnerability level of each business partner against each particular security threat scenario.

Step 2: Threat Scenario Identification

Based on the threat categorization described in the previous section, in this step, all the business partners will identify specific threat scenarios from each threat category that are relevant to the SCS under examination. At least the threat scenarios described in Table 4.4 will be examined. Each business partner may identify additional security scenarios that may not be included in Table 4.4. These new threat scenarios will also be assigned to one of the threat categories as defined in the same table.

Step 3: Likelihood Analysis

In this step, the expected likelihood for each possible threat scenario for the SCS under examination will be estimated. Each business partner will assess the likelihood of each threat scenario, taking into consideration two values: a threat value, reflecting the probability of occurrence, and a vulnerability value, reflecting the current level of protection against a threat scenario. We describe in detail the calculation of the threat, vulnerability, and likelihood level.

Threat Assessment

For each examined SCS, the probability of occurrence for each possible threat scenario is assessed. The probability of a threat scenario is defined as the expected frequency of appearance, based on the threat level scale defined in Table 4.8. When selecting the threat level for an examined threat scenario, a business partner will consider the expected probability of occurrence of the examined threat scenario within the business partner itself (i.e., the probability that the threat is initialized within its own systems and/or premises).

To assist the business partner in estimating a threat level, two selection criteria are provided in Table 4.5: (1) the probability of occurrence and (2) the history of previous incidents. The responder may select one answer for each of these criteria (possibly belonging to a different scale). If the answers provided by the business partner belong to a different scale, then the highest scale is selected, and the final outcome will be the default value corresponding to the highest value from the selected answers. Thus the methodology will always use one of the five default values, as a percentage value (Very Low = 20%, Low = 40%, Medium = 60%, High = 80%, Very High = 100%). Let $t_i(TS_j)$ denote the probability of a particular threat scenario TS_j being realized, as assessed by the business partner x_i (or the risk assessor in consensus with the business partner).

Example: Assume that for an SCS under examination and for a particular threat scenario TS_j the business partner x_i believes that the probability of occurrence of TS_j in the premises of x_i is about 50% (corresponding to the threat level "Medium"). In addition, x_i replies that TS_j has a history of one incident in the last 12 months (corresponding to the threat level "High"). Based on the answers of the business partner and Table 4.8, the threat level assigned is $t_i(TS_j)$ = High (=80%).

Vulnerability Assessment

Lack of security controls for a threat scenario will lead to a high vulnerability of this threat scenario. Vulnerability assessment will make use of the SDSs completed by each business partner (see Table 4.4), where each business partner has already provided information related with the implementation level of security controls related with the SC. In addition, the mapping of the threat categories (and consequently of threat scenarios) to the required security controls, as described in Table 4.6, can be combined to provide a complete view of the level of implementation of the security controls for each examined threat. The Medusa system transparently uses this information, provided once in the SDS, to evaluate the vulnerability level for each examined threat scenario. Each security control is in one of the three possible states: fully defined (with a score 2), partially defined (score 1), or not implemented (score 0).

In each business partner, the vulnerability level for each threat scenario is evaluated as follows: Let $v_i(TS_j)$ denote the vulnerability level of x_i for a threat scenario TS_j. For each threat, let $Impl_{max}(TS_j)$ denote the maximum possible implementation score that a business partner would get if all the proposed controls for this threat were fully implemented. Let $Impl_i(TS_j)$ denote the implementation score of the business partner x_i for the threat scenario TS_j, as computed based on Table 4.4. Then the vulnerability level of the business partner x_i for the threat scenario TS_j is computed as shown in Eq. (4.1):

$$v_i\left(TS_j\right) = \left[1 - \frac{Impl_i\left(TS_j\right)}{Impl_{max}\left(TS_j\right)}\right] \times 100 \tag{4.1}$$

The vulnerability level for each threat scenario is automatically calculated by the Medusa system for all the business partners and for all possible threat scenarios, using the completed SDSs and the assignment of threat scenarios to security vulnerabilities. Thus, the business partner does not need to perform any additional action to evaluate all the vulnerability levels.

Likelihood Assessment

The likelihood assessment for each examined threat scenario is computed as the product of the relative threat and vulnerability values. The likelihood of occurrence of a threat scenario TS_j to a business partner x_i is computed in Eq. (4.2).

$$l_i\left(TS_j\right) = t_i\left(TS_j\right) \times v_i\left(TS_j\right) \tag{4.2}$$

Step 4: Consequence Analysis

For each examined SCS, the consequence for each possible threat scenario is assessed by each business partner, using the consequence scale defined in Table 4.7. The responder may select more than one description of consequence from Table 4.7, but the final consequence level will be the value that corresponds to the highest possible consequence from the selected answers. It is important to note that when selecting a consequence level for a threat scenario, each business partner will assess the worst case consequence that the business partner may experience if the threat scenario were realized in any node within the SC (not in particular in the same business partner). Thus, the consequence assessment already considers the potential consequence of a cascading threat for the given threat scenario, and the methodology will always use one of the five default values: Very Significant (derived value 100), Significant (80), Average (60), Minor (40), and Minimal (20). Let $c_i(TS_j)$ denote the consequence of a threat scenario TS_j being realized, as assessed by the business partner x_i (or the risk assessor in consensus with the business partner).

Step 5: Risk Assessment

In this step, the likelihood and consequence values, calculated in the previous steps, will be combined to calculate the following for each threat scenario: (1) the partial risk level for each business partner and (2) the overall risk level that characterizes the whole SCS.

Partial and Overall SC Risk Assessment Evaluation

In this step, each business partner assesses the risk for each threat scenario, using as input the likelihood values and the consequence values. To calculate the final risk level, the risk scale described in Table 4.8 is used. The partial risk level of the business partner x_i for the examined threat scenario TS_j is calculated in Eq. (4.3):

$$r\left(TS_j\right) = l_i\left(TS_j\right) \times c_i\left(TS_j\right) \tag{4.3}$$

The overall risk level characterizes the risk level of the whole SCS and is calculated as follows in Eq. (4.4):

$$R\left(TS_j\right) = \max\left(w_i \times r_i\left(TS_j\right)\right), \quad \forall x_i \in \mathbb{N} \tag{4.4}$$

Here we compute the partial risk level for each business partner weighted by the business importance w_i as provided by the business partners in Step 1.1 of the methodology. The reason behind this is that the overall risk level that characterizes the whole SCS as an entity should also consider the business importance of the business partners for the provision of the service. The maximum weighted partial risk value will be the overall risk level. Fig. 4.5 shows an example of the risk assessment calculation by the Medusa system.

Step 6: Cascading Risk Assessment

In this step, the risk assessor will make use of all the risk parameters computed in the previous steps to evaluate the effect of potential cascading threats for each examined threat scenario. The cascading risk assessment evaluation is based on a modified version of the methodology proposed in [13,26]. First, all possible cascading paths are defined, based on the dependencies between the business partners within the SCG (as shown in Fig. 4.5). Then, the expected dependency risk of each examined dependency chain for each examined threat scenario is computed, based on the following algorithm:

Define a minimum and a maximum path length, ℓ_{min} and ℓ_{max}, respectively. Typical values are dependency paths between 2 and 5 nodes.

Each node within the SCG will be examined as a potential initiator of a cascading chain. Without loss of generality, say that y_0 is the source node of a chain.

For each examined source node y_0, identify all possible dependency chains that initiate from y_0 with length between ℓ_{min} and ℓ_{max}. Without loss of generality, say that $y_0 \rightarrow y_1 \rightarrow \cdots y_n$ is one dependency chain initiating from the node y_0.

For each dependency chain $y_0 \rightarrow y_1 \rightarrow \cdots y_n$ initiating from y_0 (with length between ℓ_{min} and ℓ_{max}), compute the cumulative cascading dependency risk $r_{01\ldots n}(TS_j)$ as illustrated in Eq. (4.5):

$$r_{01\ldots n}\left(TS_j\right) = \left(\prod_{k=0}^{i} l_k\left(TS_j\right)\right) \cdot c_i\left(TS_j\right) \tag{4.5}$$

The likelihood product, defined in Eq. (4.5), is a recursive ordered operation, where the leftmost (first) pair always has the highest priority. The product of the likelihood values, defined in Eq. (4.5), is computed based on Table 4.9.

Note that Table 4.9 is constructed in such a way that the product of likelihood operation is commutative. However, we note that since the product of likelihood operator is ordered, the operation is not associative. By using an ordered operation to compute the product of likelihood in the cascading risk calculation, we achieve the following properties.

The likelihood value of the source node of the dependency chain has the lowest effect on the outcome, while the likelihood value assessed by the destination of the dependency chain always has the highest weight in the result.

Table 4.9 Product of likelihood values calculation

	VL	L	M	H	VH
VL	VL	VL	L	L	M
L	VL	L	L	M	H
M	L	L	M	H	H
H	L	M	H	H	VH
VH	M	H	H	VH	VH

Table 4.10 An example of input values for the calculation of cascading risk

	x_1	x_2	x_3
l_i	VL −20%	H −80%	L −20%
c_i	Average (60)	Significant (80)	Minor (40)
r_i	VL	H	VH

As a result, Eq. (4.5) succeeds to capture the distance of a business partner from the source of the event. Due to the ordered definition of the operator, the effect of the most distant nodes (the nodes having dependencies of higher order with the destination of the examined cascading threat) will "fade off" faster than the effect on the nodes that are closer to the examined destination (target) of the cascade threat. The following example clarifies these properties.

Example: Assume that for an SCS under examination, the likelihood, consequence, and partial risk values shown in Table 4.10 hold. Let us assume that the SCG in question includes the dependency chain: $x_1 \rightarrow x_2 \rightarrow x_3$.

Now, by using Eq. (4.5), we will compute all the dependency risk values for node x_3.

Calculation of the first-order dependency risk value(s) for node x_3: Based on the examined chain $x_1 \rightarrow x_2 \rightarrow x_3$, we have one such first-order dependency for the node x_3, and we compute (we omit the (TS_j) for simplicity):

$$r_{23}\left(TS_j\right) = (l_2 \cdot l_3) \times c_3 = (H \cdot L) \times \text{Minor} = M \times \text{Minor} = \text{Low}.$$

Interestingly, the first-order dependency risk of x_3, r_{23} = Low, which is higher than the partial risk r_3 (=Very Low). Calculation of the second-order dependency risk value(s) for node x_3: Based on the examined chain $x_1 \rightarrow x_2 \rightarrow x_3$, we have one such second-order dependency for the node x_3, and we compute:

$$\begin{aligned} r_{123}(TS_j) &= (l_1 \cdot l_2 \cdot l_3) \times c_3(TS_j) \\ &= ((VL \cdot H) \cdot VH) \times \text{Minor} \\ &= (L \cdot VH) \times \text{Minor} \\ &= H \times \text{Minor} \\ &= \text{Low Risk.} \end{aligned}$$

Note that the effect of the rightmost likelihood value (i.e., the likelihood assessed by the destination of the cascading chain) in the final product of the likelihood is higher than the effect of the leftmost value (the source of the chain). Indeed, observe that in the previous example, if the operator was computed with a rightmost priority, the estimated likelihood would be different, $((VL \cdot (H \cdot VH)) = VL \cdot VH = M)$, ultimately resulting in a higher cascading risk. The definition of the product of likelihood operator as a leftmost ordered operation favors the likelihood values of the nodes having smaller distance (order of dependency) to the destination of the cascade risk value, as required by the methodology. Fig. 4.2 shows the general architecture of the Medusa system that the Medusa methodology has been implemented.

Remark: The cascading dependency risk captures the risk that a business partner may experience within the SC due to a (related) threat scenario being realized in another business partner from whom the business partner in question has a dependency. Recall that when providing the consequence level, each business partner considers the worst case impact for itself, regardless of the initiator of the threat scenario that may cause the consequence to happen. On the other hand, the likelihood value computed for each business partner only considers that the business in question will initiate the threat scenario. The cascading dependency risk captures the cascading effect of a threat scenario realized in some business partner to affect others due to their (direct or indirect) dependencies from the source of the event.

Step 7: Risk Mitigation

This step of the methodology aims to assure an acceptable risk level for the whole SCS. The basic idea is to calculate the risk level of each business partner in the supply chain for every possible threat. In case there are some business partners that do not satisfy the desired risk level (acceptable threshold), the methodology needs to determine which security controls must be implemented and which partners need to implement them to resolve these issues.

We break the methodology into two algorithms: the first is for the supply chain as a whole, and the second is for the cascading dependency chains. Before we present the algorithms, we will introduce some terminology.

Recall the partial risk level $r_i(TS_j)$ of business partner x_i for threat scenario TS_j will be used as a metric for the prioritization of possible additional security controls. Furthermore, recall the overall risk level, $R(TS_j)$, characterizing the risk level of the whole supply chain, will be used as a metric to define a baseline risk level (the minimum acceptable security level) for anyone participating in the SCS. The goal will be to reduce the risk level $r_i(TS_j)$ of every business partner (and consequently $R(TS_j)$ below the risk threshold) by applying additional security controls.

THE MEDUSA SCS RISK ASSESSMENT SYSTEM

The Medusa system implements the Medusa methodology and aims to systematically evaluate the security risks affecting the supply chain business partners within an SCS. Fig. 4.2 depicts the overall architecture of the Medusa system, which consists of four main conceptual layers: Users, Information Assets, Components, and Technological Infrastructure.

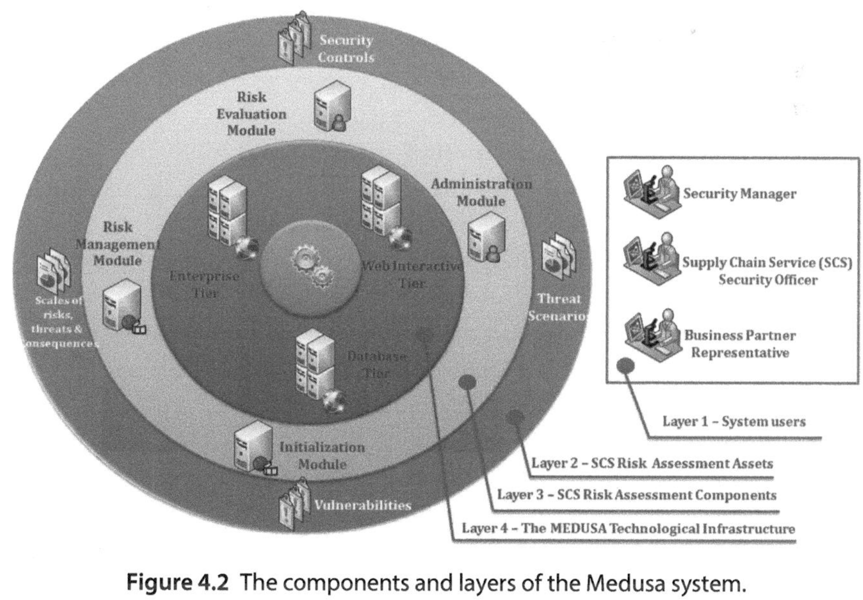

Figure 4.2 The components and layers of the Medusa system.

These layers constitute the conceptual pillars for building and implementing the ICT tools that will support the modeling and visualization of the various security-related risks and interdependencies. More specifically:

Layer 1 – System Users: The Medusa ecosystem consists of the following user groups:

Security Manager: The purpose of the Security Manager is to initialize the system with information that is globally available to the other two roles. In particular, he/she has to refine and populate the closed "closed lists"/vocabularies that will be used by the other layers.

SCS Security Officer: He/she is responsible for the initiation of a risk assessment and the specification of the SCS that will be examined identifying all the involved business partners and the supported processes.

Business Partner Representative: This is the representative of a business partner (e.g., customs, insurance company, shipping company, maritime ministries, public authorities) that participates in the SCS evaluation process and provides information relating to the threats, vulnerability, and controls that belong to the area of his/her responsibility.

These user groups are defined according to their roles, requirements, and responsibilities; therefore, they have different access rights to the components and tools of the system.

Layer 2 – SCS Risk Assessment Information Assets: All user groups provide initial content to the system including but not limited to the following: threat scenarios, security controls, vulnerabilities, and types of enumerations such as types of dependencies (e.g., access to cyber systems and e-data, access to physical facilities), types of business partners (e.g., ports, customs, importers, exporters, ship owners), categories of consequences and scale of risks, threats, and consequences (e.g., high).

Layer 3 – SCS Risk Assessment Components: All SCS risk assessment assets are accessed, properly managed, and processed by the implemented components. These components are as follows:

Administration module allows customizing the parameters, elements, and features (e.g., threats, vulnerabilities, controls) required for the execution of a risk assessment process based on the defined Medusa methodology.

Initialization module allows the initiation of a risk assessment and the definition of the SCS that will be examined.

Risk Evaluation module gives the opportunity to the users to assess the partial risk level of the involved business partners for the examined threat scenarios partial risk of the examined SCS.

Risk Management module allows the review of the risk assessment results and the calculation of the cumulative cascading dependency risk values for the applicable threat scenarios.

Layer 4 – The Medusa Technological Infrastructure: This includes all the different modules, subsystems, and primary services that comprise the Medusa system.

In this context, the Medusa system models and implements the main steps of the proposed Medusa Supply Chain Risk Assessment Methodology, described in Medusa: A Maritime SCS Risk Assessment Methodology section, in a graphic way using visualization tools, simulation processes, automated routines, and structured content. The adopted approach encapsulates several simplifications and incorporates automated steps to deliver and provide automated analysis and evaluation of all activities comprising an SCS, taking into consideration their complexity and distributed nature.

As shown in Fig. 4.3, the Medusa system transparently uses the implementation level of the relevant security controls, provided once in the SDS, to evaluate the vulnerability level for each examined threat scenario (Step 1.2).

In Fig. 4.4, all threat scenarios that are relevant to the Vehicles Transport SCSs are evaluated by estimating their probability of occurrence (Step 3).

Fig. 4.5 shows the overall risk levels of the Vehicles Transport SCSs for all examined threat scenarios as calculated by the Medusa system (Step 5). Threat scenarios are ranked according to their derived risk level.

Finally, Fig. 4.6 shows all possible dependency chains and presents the corresponding cascading dependency risk values (Step 6).

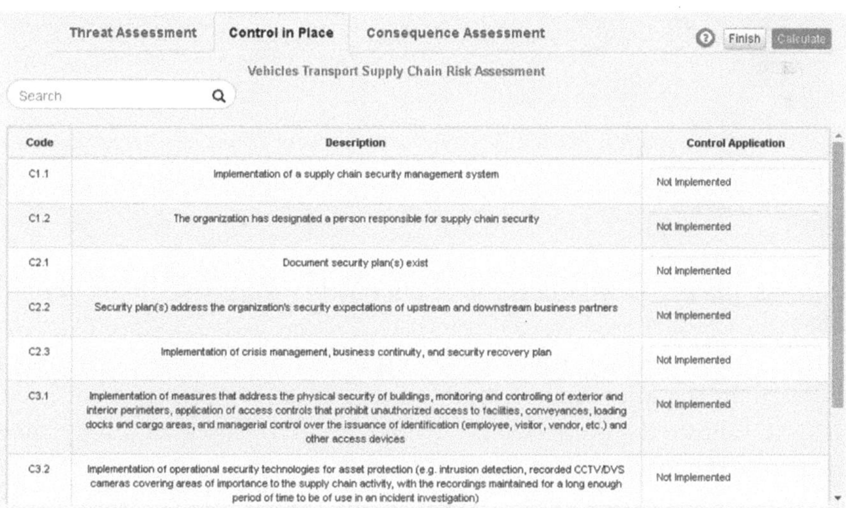

Figure 4.3 Using the control implementation level, as provided in the Security Declaration Statement, to assess the vulnerability level for each threat in the Medusa system.

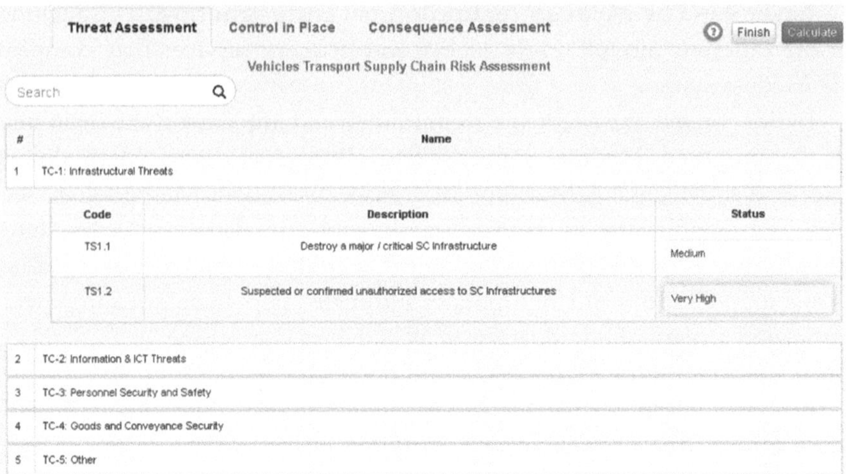

Figure 4.4 Assessment of each threat scenario that is relevant to the vehicle transport SCS in the Medusa system.

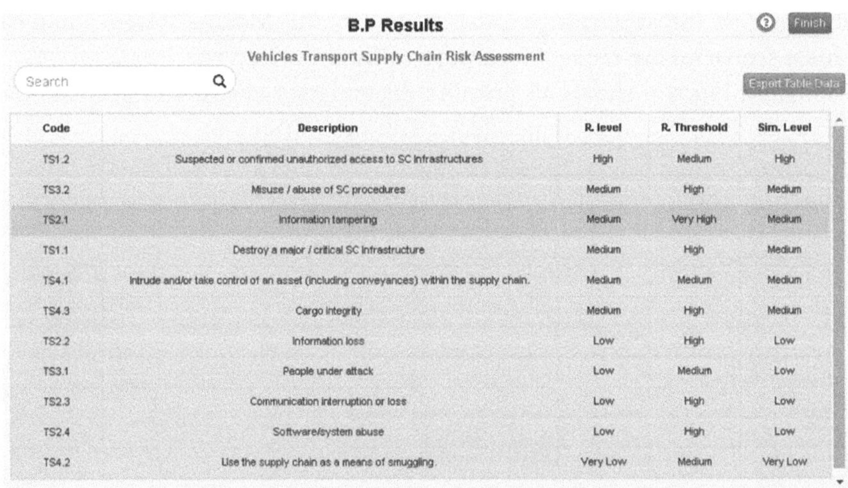

Figure 4.5 The risk assessment phase of the Vehicle Transport SCS in the Medusa system.

Based on the pilot tests of the Medusa project, we used reference scenarios to validate the proposed methodology and the integrated algorithms in terms of the following:

validating the evaluation of the partial and overall risks;

validating the evaluation of the multiorder dependencies and risks;

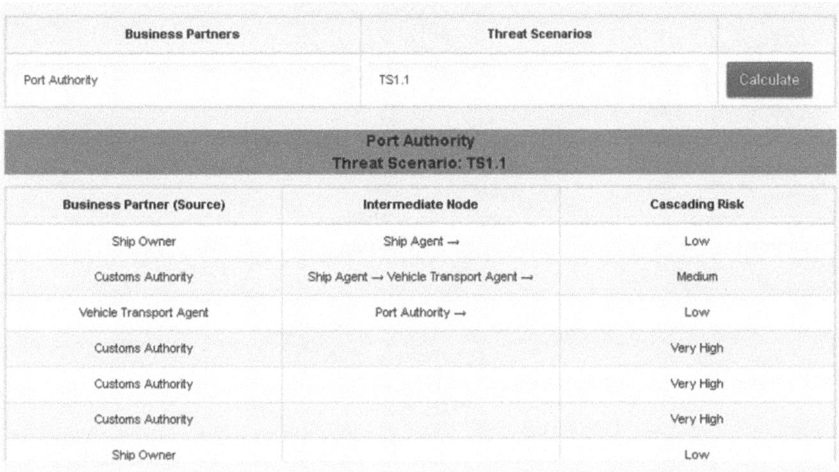

Business Partners	Threat Scenarios	
Port Authority	TS1.1	Calculate

Port Authority
Threat Scenario: TS1.1

Business Partner (Source)	Intermediate Node	Cascading Risk
Ship Owner	Ship Agent →	Low
Customs Authority	Ship Agent → Vehicle Transport Agent →	Medium
Vehicle Transport Agent	Port Authority →	Low
Customs Authority		Very High
Customs Authority		Very High
Customs Authority		Very High
Ship Owner		Low

Figure 4.6 The cascading dependency risk assessment phase by the Medusa system.

validating the consistency of the results by escalating the values of all the parameters of the methodology.

We describe, analyze, and interpret the results of the experimental validation process.

Test Scenario: The "Vehicles' Transport" Supply Chain Service

To validate the accuracy, robustness, efficiency, and reliability of the Medusa risk management methodology, we will use the *Purchase and Shipment* supply chain subprocess of the "Vehicles Transport Service." This process is initiated by the *Importer*, which sends a purchase order to the *Industry* for several vehicles. Once both have agreed upon the terms of the contract (pricing, documentation, freight charges, currency, etc.), the Industry contracts a *Ship Agent* to deliver the vehicles to the destination port. The *Ship Agent* makes the arrangements with the *Ship Owner* to assure usage of ships, with *Customs Authorities* to arrange for the Manifest Registration Number and the departure *Port Authority* to arrange the ship formalities related to the authorization process from the entry of the ship into the port until its exit and then proceed to load the vehicles into the vessel for shipment to the destination port. The Ship Agent contracts a *Vehicle Transport Agent* and assigns the transfer of the vehicles from the Industry to the departure port. Finally, the Ship Agent sends the relevant documentation to the Importer's *Local Agent*, who has the responsibility for the ship arrival and the regional procedure of delivering the

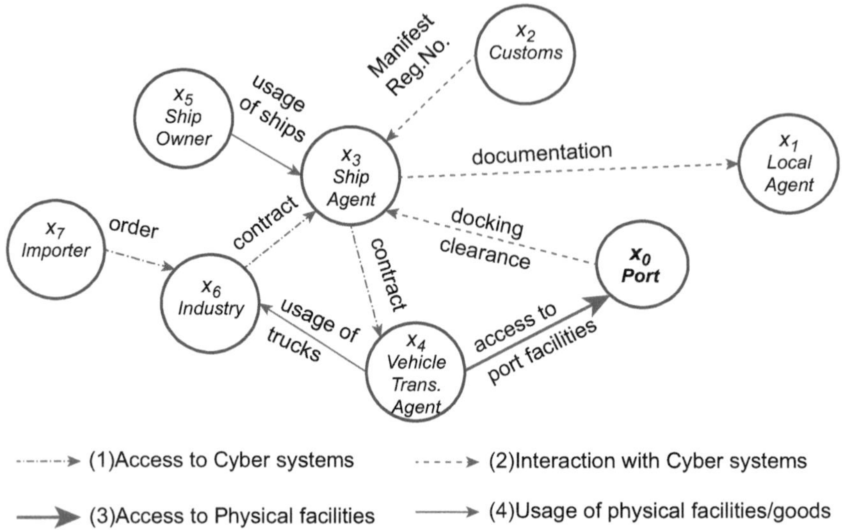

Figure 4.7 An SCG based on the dependencies of the Purchase and Shipment SC.

vessel to the importer. Fig. 4.7 shows the SCG representing the dependencies between all the aforementioned entities (cyber or physical ones).

VALIDATION SCENARIOS

Within the context of the experimental validation process of the aforementioned part of the "Vehicles Transport Service," the following scenarios will be examined:

Validation Scenario 1: implementation of different combinations of security controls

Validation Scenario 2: variation of the expected consequences

Validation Scenario 3: variation of the likelihood of occurrence of threat scenarios

Validation Scenario 4: overall and cascading dependency risks validation

Validation Scenario 1: Implementation of Different Combinations of Security Controls

In the first validation scenario, for the Purchase and Shipment SC subprocess, we will examine the threat $TS_{1.1}$ (destroy a major/critical SC infrastructure); in particular, we will explore how the risk value of $TS_{1.1}$ changes if a business partner (i.e., port authority, x_0) applies different mitigations controls. It should be noted that the vulnerability level of the port (x_0) for the threat scenario $TS_{1.1}$ (as well as for any examined threat) is automatically calculated based on the implementation level of security controls related with it. Thus each business partner will provide *once* all the required input related with the implementation level of its security controls, and the vulnerability level of each threat is automatically calculated. For the examined scenario, the relevant controls are $C_{1.1}$, $C_{1.2}$, $C_{2.1}$, $C_{2.2}$, $C_{2.3}$, $C_{3.1}$, $C_{3.2}$, $C_{3.3}$, and $C_{3.4}$. To validate the consistency of the methodology, we will examine the behavior of Medusa assuming there are in place various combinations of security controls. Each combination of implemented security controls simulates a different security implementation maturity level for the related security controls. We examine the following four different combinations of security controls for the port authority:

Security Controls Combination A (SCC_A): All the proposed controls related with the examined threat are fully implemented.

Security Controls Combination B (SCC_B): All the proposed controls related with the examined threat are partially implemented.

Security Controls Combination C (SCC_C): One-third of the proposed controls are fully implemented, one-third is partially implemented, and one-third of the related security controls are not implemented.

Security Controls Combination D (SCC_D): None of the proposed controls related with the examined threat are implemented.

Notice that these combinations of security controls implementation attempt to capture both extreme cases (all controls are implemented in SCC_A, no controls are implemented in SCC_D), as well as intermediate scenarios of control implementation maturity. The combination SCC_C would be more realistic, since the high-level controls are first assumed as fully implemented, and as the controls become more low-level, the implementation level decreases. Table 4.11 presents the case scenarios that have been executed.

Case Scenario 1.a: In the first case, we assume that the probability of occurrence of threat scenario $\mathbf{TS_{1.1}}$ is more than 80%, and its expected consequences are "Very Significant." The results as derived from the preceding combinations of security controls show that if the expected vulnerability level between two cases is equal, despite the exact combination of implemented controls (cases SCC_B and SCC_C), then the resulting risk level is the same. Intermediate scenarios were also tested as subcases and also proved consistency of the tool. For example, if the combination of control implementation gives low vulnerability level (e.g., 20%), then the expected risk will be "Low."

Case Scenario 1.b: Assuming that the probability of occurrence of threat scenario $\mathbf{TS_{1.1}}$ is more than 80% and the consequences that the

Table 4.11 Validation scenario 1: implementation of different security controls and related risk levels
Risk parameters for port (x_0) for $TS_{1.1}$

Case scenario	**Case scenario 1.a**				**Case scenario 1.b**			
Security controls combinations	SCC_A	SCC_B	SCC_C	SCC_D	SCC_A	SCC_B	SCC_C	SCC_D
Threat level	Very high				Very high			
Vulnerability level	0%	50%	50%	100%	100%	50%	50%	0%
Likelihood level	Very low	Low	Low	Very high	Very high	Low	Low	Very low
Consequence level	Very significant				Minimal			
Risk level	Very low	Medium	Medium	Very high	Very low	Very low	Very low	Very low

organization would experience is "Minimal," we can infer that a threat scenario with an expected minimal consequence level will always output a very low risk, despite the other parameters (e.g., implemented controls).

Case Scenario 1.c: If the probability of occurrence of this scenario is about 10% (corresponding to the threat level "Very Low") and the consequences of the particular threat scenario being realized are "Very Significant," then taking account the preceding Security Controls Combinations, we can say that a threat scenario with an expected very low probability of occurrence will always output a very low risk, despite the other parameters.

Case Scenario 1.d: In this case, we consider that the probability of occurrence of this scenario is about "Medium" and the possible consequences are "Very Significant." The derived results show that a threat scenario of very high significance will exhibit a risk level depending on its likelihood of occurrence (which is calculated based on the threat and vulnerability levels).

Validation Scenario 2: Variation of the Expected Consequences

In the second scenario, we will gradually change the level of the expected consequences of a threat scenario $TS_{1.1}$ while maintaining the same values for the threat and vulnerability level to inspect how the corresponding risk value is affected. At this point, we will investigate how the variations in the consequence affect the risk level. In this context, we assume that the potential consequences of the particular threat scenario vary as follows:

Consequence Scenario A (CS_A): The consequences that the organization would experience are "Very Significant."

	Case scenario 1.c				Case scenario 1.d			
	SCC_A	SCC_B	SCC_C	SCC_D	SCC_A	SCC_B	SCC_C	SCC_D
	Very low				Medium			
	0%	50%	50%	100%	0%	50%	50%	100%
	Very low	Very low	Very low	Very low	Very low	Low	Low	Medium
	Very significant				Significant			
	Very low	Very low	Very low	Very low	Very low	Low	Low	Medium

Consequence Scenario B (CS_B): The consequences that the organization would experience are "Significant."

Consequence Scenario C (CS_C): The consequences that the organization would experience are "Average."

Consequence Scenario D (CS_D): The consequences that the organization would experience are "Minor."

Consequence Scenario E (CS_E): The consequences that the organization would experience are "Minimal."

To this end, the following five main cases have been examined (results shown in Table 4.12):

Case Scenario 2.a: In this case, we make the following assumptions for the port (x_0): (1) the probability of occurrence of the threat $TS_{1.1}$ at the port's premises is more than 80% (corresponding to the threat level "Very High"); and (2) the port applies a combination of security controls that are based on the vulnerability level formula defined in methodology, so the vulnerability level of the port authority of the examined threat scenario is rather high (72%). This combination leads to a "High Likelihood" value for all the examined cases.

Case Scenario 2.b: The assumptions made for the port (x_0) are the following: (1) the examined threat scenario $TS_{1.1}$ is "Very Likely" to occur (corresponding to the threat level "Very High"); and (2) the port does not apply any security controls related to this threat; therefore, based on the defined score the vulnerability level of the port is very high (100%). This combination leads to a "Very High Likelihood" value for all the examined cases.

Case Scenario 2.c: The assumptions made for the port (x_0) are the following: (1) the probability of occurrence of the examined threat scenario

Table 4.12 Validation Scenario 2: variation of the expected consequences and related risk levels
Risk parameters for port (x_0) for $TS_{1.1}$

Case scenario	**Case scenario 2.a**					**Case scenario 2.b**					**Case scenario 2.c**	
Consequence scenario	*CS_A*	*CS_B*	*CS_C*	*CS_D*	*CS_E*	*CS_A*	*CS_B*	*CS_C*	*CS_D*	*CS_E*	*CS_A*	*CS_B*
Threat level	Very high					Very high					Very high	
Vulnerability	72%					100%					50%	
Likelihood	High					Very high					Medium	
Consequence	VS	S	A	Minor	Minimal	VS	S	A	Minor	Minimal	VS	S
Risk level	H	M	L	L	VL	VH	H	M	L	VL	M	L

$TS_{1.1}$ is "High"; (2) the port partially applies all the security controls related to this threat. This combination leads to a "Medium Likelihood" value for all the examined cases.

Case Scenario 2.d: In this scenario, for the port (x_0), we assume: (1) the examined threat scenario $TS_{1.1}$ has a medium probability of occurrence (corresponding to the threat level "Medium"); (2) the port applies a set of security controls that stands for a low-to-medium (38%) vulnerability level. This combination leads to a "Low Likelihood" value for all the examined cases.

Case Scenario 2.e: In last case, (1) the examined threat scenario $TS_{1.1}$ has "Low" probability of occurrence (corresponding to the "Low" Threat Level); the port (x_0) implements few of the proposed security controls for this threat, so the calculated vulnerability level for the examined threat is high (83%).

The results of the validation scenario 2 showed, for all the subcases (2.a–2.e), that the risk values for all cases were as expected and consistent with the risk calculation process of the Medusa methodology. As a consequence, the value varies, while all the other parameters are not changed, and the derived risk depends on the variation of the consequence level. The subcases capture various cases of threat–vulnerability combinations, where in each subcase the derived likelihood value is the same.

Validation Scenario 3: Variation of the Probability of Occurrence of Threat Scenarios

The main objective of the third validation scenario is to investigate how the risk of a threat occurring varies as the threat level (probability of occurrence)

			Case scenario 2.d					**Case scenario 2.e**				
CS_C	CS_D	CS_E	CS_A	CS_B	CS_C	CS_D	CS_E	CS_A	CS_B	CS_C	CS_D	CS_E
			Medium					Low				
			38%					83%				
			Low					Very low				
A	Minor	Minimal	VS	S	A	Minor	Minimal	VS	S	A	Minor	Minimal
L	L	VL	L	L	L	VL	VL	L	L	L	VL	VL

of the particular threat varies. In this context, we study the following five frequency scenarios (see Table 4.13):

Frequency Scenario A (*FS_A*): The examined threat scenario is "Very Likely" to occur.

Frequency Scenario B (*FS_B*): The examined threat scenario is "Likely" to occur.

Frequency Scenario C (*FS_C*): The examined threat scenario has "Medium Likelihood" to occur.

Frequency Scenario D (*FS_D*): The examined threat scenario is "Unlikely" to occur.

Frequency Scenario E (*FS_E*): The examined threat scenario is "Very Unlikely" to occur.

Case Scenario 3.a: In the first case the following assumptions are made: (1) the consequences that the port (x_0) would experience due to $TS_{1.1}$ being realized are "Very Significant"; and (2) the port (x_0) implements a set of security controls related with this threat; based on this implementation the derived vulnerability level is medium (45%).

The results of this case show that a threat scenario of very high significance will exhibit a risk level depending on its likelihood of occurrence (which is calculated based on the threat and vulnerability levels). By maintaining the vulnerability level at a constant value and by varying the threat level from "Very Low" to "Very High," we can verify that the resulting risk level is gradually increasing based on the derived likelihood level.

Case Scenario 3.b: We assume that (1) the consequences that the port (x_0) would experience due to $TS_{1.1}$ being realized are "Minimal"; and (2) the

Table 4.13 Validation Scenario 3: variation of the probability of occurrence of the threat scenarios and related risk levels
Risk parameters for port (x_0) for $TS_{1.1}$

Case scenario	Case scenario 3.a					Case scenario 3.b				
Frequency scenario	*FS_A*	*FS_B*	*FS_C*	*FS_D*	*FS_E*	*FS_A*	*FS_B*	*FS_C*	*FS_D*	*FS_E*
Threat level t()	VL	L	M	H	VH	VL	L	M	H	VH
Vulnerability v()	45%					72%				
Likelihood level l()	VL	L	L	L	M	Vl	VL	L	M	M
Consequence value c()	Very significant					Minimal				
Risk level r()	VL	L	L	L	M	VL	VL	VL	VL	VL

port (x_0) implements only a few of the proposed controls for this threat; thus, the corresponding vulnerability level is "High" (72%)

The results deduce that a threat scenario of very low significance ("Minimal" consequence) will always exhibit a very low risk level regardless of the threat level.

Case Scenario 3.c: In this case, we assume that (1) the possible consequences of the $TS_{1.1}$ to the port (x_0) are "Average"; and (2) the port (x_0) implements partially all the proposed controls for this threat, deriving a vulnerability level of 50%.

The results show that a threat scenario with an expected medium significance ("Average" consequence) with a medium vulnerability level will at worst exhibit a "Low" risk level regardless of the threat level.

Case Scenario 3.d: We made the following assumptions: (1) the possible consequences of the $TS_{1.1}$ to the port (x_0) are "Average"; and (2) the port (x_0) implements most of the proposed controls for this threat, deriving a very low vulnerability level of 17%.

Based on the results of this case, a threat scenario with a "Very High" consequence level and with a very low vulnerability level will at worst exhibit a "Low" risk level regardless of the threat level. The variance of the threat level influences the transience from the "Very Low" to the "Low" risk level, when the threat level is expected to be high or very high.

Validation Scenario 4: Cascading Dependency Risks Validation

In the preceding sections, various scenarios validating the risk assessment in a single business partner were presented. Those scenarios covered all the parameters of the methodology used in risk assessment (threat, vulnerability,

Case scenario 3.c					Case scenario 3.d				
FS_A	FS_B	FS_C	FS_D	FS_E	FS_A	FS_B	FS_C	FS_D	FS_E
VL	L	M	H	VH	VL	L	M	H	VH
50%					17%				
Vl	VL	L	M	M	VL	VL	VL	L	L
Average					Very significant				
VL	L	L	L	L	VL	VL	VL	L	L

and consequence levels). This section presents a validation scenario covering the cascading dependency effects associated with the *Purchase and Shipment* SC subprocess of the "Vehicles Transport Service."

In particular, we use the threat scenario $TS_{1.1}$ (destroy a major/critical SC infrastructure) to examine the cascading aspects of the proposed methodology. Assuming that the threat and consequence values for $TS_{1.1}$ shown in Table 4.14 were provided by each business partner and that the vulnerability values were computed by applying Eq. (4.5) on the input provided by the SDS of each partner, then the partial risk values for this threat scenario are calculated.

Based on the preceding values, we examined the cascading dependency risk values for the following case:

Dependency scenario: The port authority is the destination of the cascading threat. If the port authority is the destination of the cascading threat, then the following dependency chains are produced (Table 4.15):

The results of this scenario verify that the outcome is consistent with the Medusa cascade risk calculation. We notice that based on the input values, the port authority (along with the customs authority) were the nodes with the higher partial risk for the examined threat with a "Medium" threat value. If, however the threat originates near the port authority to its first-order or second-order incoming dependencies (i.e., the Vehicle Transport Agent or the Ship Agent) then the Incoming Dependency Risk for the port authority is still considered "Medium."

This implies that if additional security controls should be implemented, these should be implemented either at the port authority or at its first-order dependencies. As the initiator of the threat event is more distant than the

Table 4.14 Validation Scenario 4: cascading risks

Risk parameters for all business partners for $TS_{1.1}$					
Business partners	**Threat level**	**Vulnerability**	**Likelihood level**	**Consequence value**	**Risk level**
Port (x_0)	VH	L (38%)	L	Very significant	Medium
Local agent (x_1)	L	M (50%)	VL	Average	Very low
Customs (x_2)	M	M (45%)	L	Very significant	Medium
Ship agent (x_3)	M	M (60%)	L	Significant	Low
Vehicle transp. agency (x_4)	H	H (72%)	M	Low	Low
Ship owner (x_5)	M	L (36%)	L	Very significant	Low
Vehicle industry (x_6)	M	L (22%)	L	Significant	Very low
Importer (x_7)	L	H (77%)	L	Average	Low

Table 4.15 Dependency chains with the port authority as the destination for the threat scenario $TS_{1.1}$

Dependency chain	Dependency risk
Vehicle transport agent→port authority	Medium
Ship agent→vehicle transport agent→port authority	Medium
Industry→ship agent→vehicle transport agent→port authority	Low
Ship agent→vehicle transport agent→port authority	Low
Customs→ship agent→vehicle transport agent→port authority	Low
Industry→ship agent→port authority	Low
Ship owner→ship agent→vehicle transport agent→port authority	Low
Ship owner→ship agent→port authority	Low
Customs authority→ship agent→port authority	Low

examined destination (the port authority), then the dependency risk for the port authority gradually decreases.

Medusa attempts to cover two main goals: First it aims to be a practical SC risk assessment methodology fully compliant with ISO 28001. In addition, Medusa aims to capture and assess the cascading risks within the supply chain. Toward this, Medusa extends ISO 28001 in several ways. First, it defines threat categories and uses them to group specific threat scenarios. Then the threat categories are mapped to specific subsets of the security controls already defined in the SDSs. Thus by using the statements provided by all business partners, it is easy to quantify the vulnerability (and eventually likelihood) of each partner for each examined threat scenario. In addition, Medusa allows the efficient assessment of cascading risks within the SC by using input data already provided in the previous steps; all potential dependency chains of the SC are assessed, and their cascading risk values are then used to prioritize risk mitigation. The Medusa methodology is characterized by its capabilities to capture the opinions, experience, and expertise of the business partners engaged in the risk assessment process and produce solid results; be resistant against outliers and model deviations by robust estimation of the risks; provide more objective risk estimates for the SCS since the opinions of all involved business partners are considered; and provide results that are robust to variations in model parameters.

The Medusa methodology is implemented in the Medusa system offering SC risk assessment services to the business partners of the SCS, which is not limited to the borders of one organization, a single country, or region. However, further research and collaboration is required in the security of International Supply Chain Services (ISCS) with harmonized implementation of the existing international rules (e.g., ISPS, EC Regulations No 725/2004, 2005/65, 2013/0027, 2015 European Parliament Proposed Directive DATAPROTECT 235, and NIS/2015). New modeling techniques for visualizing and analyzing the complex interdependencies of physical and cyber SC assets, innovative risk assessment methods, and tools for ISCS identifying and forecasting propagated vulnerabilities and attack trees are emerging research directions.

Finally, it should be noted that during the pilot operations of the Medusa project, many port's operators, maritime stakeholders, and security/safety experts have been engaged in the process of evaluating the capacity of the Medusa methodology and system (http://medusascsra.cs.unipi.gr/) to meet their objectives. In particular, more than 400 port operators, government officials, leading experts from the maritime, oil, and gas sector, and IT security professionals trained on the functionality and services of the Medusa system (including representatives from Valencia port Foundation, Port Authorities of Alicante, and Castellon and Piraeus Port Authority), and about 123 of them have used the system to identify and assess the threat scenarios and risks associated with the SCSs in which their organization participates.

CHAPTER 5

Maritime Supply Chain Risk Assessment (at Asset Level)

A plethora of distributed ICT assets from these various supply chain (SC) business partners interrelate and interact during the provision of a supply chain service (SCS) (see Chapter 2); their isolated threats and vulnerabilities may propagate in the network, putting in danger the whole SC.

In this context, the lack of visibility and traceability in the often opaque processes and practices used to develop and acquire ICT-related products and services from each actor increases the risk of not being able to detect and remediate intentional and unintentional compromise that may be introduced through a variety of means, including counterfeit materials and malicious software. In an interconnected, automated electronic age, the modern SC itself is ICT enabled, so its processes should also be protected from disclosure and exploitation. Overall, there is a clear need for targeting sophisticated global risk assessment frameworks to deal with the cascading effects risks, threats, and vulnerabilities associated with the ICT-based logistics and related SCs.

STANDARDS AND METHODS

Novel, integrated, collaborative SC risk assessment methodologies are needed, which will enable the involved SCS business partners to collaborate in the identification and classification of the various risks in the SCS interrelated cyber assets, while facilitating them in risk resolution and the creation of related SC plans.

There is a family of ISO standards addressing the security management of the SCs:

- The ISO 28000:2007 [24] is a management system specification that has been developed and introduced in response to a demand from the transportation and logistics industry for a common security management standard, with the ultimate objective of improving the overall security of SCs.
- ISO 28001:2007 [23] provides specific guidance for implementation of a security management system for the SC. It is intended to assist organizations in establishing reasonable levels of security and making better risk-based decisions for protection of the SC.

Port Cybersecurity
ISBN 978-0-12-811818-4
http://dx.doi.org/10.1016/B978-0-12-811818-4.00005-8

- The ISO/IEC 27005:2008 [26] is a commercial standard that describes the risk management process and its activities for information security. Further, it provides guidelines for information security risk management and supports the general concepts specified in ISO/IEC 27001:2005 [27] as well as the main principles and rules described in ISO/IEC 27002:2005 [28].

Hence, the objectives of these standards are not intended to constitute a risk management method but, rather, to establish a comprehensive framework and to describe requirements for the risk assessment process, for the identification of the threats and vulnerabilities, allowing one to estimate the risks in advance, their level, and finally to define an effective treatment plan. The generic nature of these standards does not include aspects that promote the collaboration among the users. Although efforts have been made to standardize SC security risk assessment, there is a lack of targeted methodologies. Medusa (see Chapter 4) developed a novel risk assessment methodology, compliant with ISO28001, ISO27001, and ISPS. However, it can be used to assess the risks of a SCS service at the entity level. The derived overall risk values are used to generate a baseline SCS security policy, identifying the least necessary security controls for each participant in the SCS. In addition, Medusa assesses the risk of cascading threat scenarios within an SCS. This enables the SCS participants to adapt their security policies according to their business role as well as their dependencies.

The Medusa methodology concentrates on the business needs and requirements of the maritime SCs. The Medusa tool that implements the methodology enables port operators and SC business partners to conduct risk assessments of the SC services that are involved in each of their domains.

The industry standard for information security vulnerabilities that can be used for real-time risk assessment at asset level is the *Common Vulnerabilities and Exposures (CVE)*: Requirements and Recommendations for CVE Compatibility [22]. It is a central repository that provides information for publicly known security vulnerabilities and exposures. This initiative runs a program named "Common Vulnerabilities and Exposures Compatibility Program" [CVE Compatibility], which has defined a list of requirements and recommendations against which any tool, service, website, database, or system that uses the CVE-related information can be evaluated to declare if they are CVE Compatible.

The main requirements and recommendations fall into the following categories:

- High-level requirements: These are the high-level requirements for all capabilities. Many of them are described in detail in later sections.

- Accuracy: CVE Compatibility only facilitates data sharing if the capability's mapping is accurate. Therefore, CVE-Compatible capabilities must meet minimum accuracy requirements.
- Documentation: The following requirements apply to documentation that is provided with the capability.
- CVE date usage: Users must know how "up-to-date" a capability's repository is with respect to its mapping to CVE. The capability owner needs to indicate the currency of a mapping by providing the date of its last update of CVE information and indicate what portion of CVE content they utilize and from where they gather the CVE content.
- Different styles of CVE name support: A capability MUST function with CVE names independent of the format of the CVE name's representation in the capability, whether it is using the older style four-digit CVE ID Syntax or the four-digit or higher digit CVE ID Syntax (used after the CVE ID Syntax modification in use after December 31, 2013).

The EU project CORE [25] project uses the aforementioned standard to develop an innovative approach to design global SCs resilient (in real time) to major disturbances caused by high-impact events. However, CORE does not provide a concrete risk assessment methodology for the global SCS.

Another industry standard that can be used in real-time risk assessment at asset level is the *Common Attack Pattern Enumeration and Classification (CAPEC)* [20], which provides a taxonomy of known threats and attacks that can be used to better understand their impact and to facilitate their mitigation. The Compatibility Program [CAPEC Compatibility] of this initiative has specified rules and requirements that a product, service, or systems must fulfill to be registered as "CAPEC-Compatible." The CAPEC main requirements and recommendations have been grouped as follows:

- High-level requirements: The following items define the concepts, roles, and responsibilities related to the proper use of CAPEC identifiers to share data across separate security analysis, security testing, security operations, and security management capabilities (tools, repositories, services, and standards) to allow these capabilities to be used together, and to facilitate the comparison of security-relevant tools and services.
- Accuracy: CAPEC Compatibility only facilitates data sharing and correlation if the capability's mapping is accurate. Therefore, CAPEC-Compatible capabilities must meet the following minimum accuracy requirements.

- Documentation: The following requirements apply to documentation that is provided with the capability.
- CAPEC version usage: Users must know what version of CAPEC is used in a capability's repository with respect to its mapping to CAPEC. The capability owner can indicate the currency of a mapping by referencing the relevant CAPEC version and, optionally, the date the mapping was updated.

The EU project MITIGATE [92] provides a radical shift in SCS risk management methodologies toward a collaborative, evidence-driven, real-time approach (MITIGATE methodology) that alleviates the limitations of state-of-the-art risk management frameworks and meets the requirements of the latter two industry standards. The project integrates, validates, and commercially exploits an effective, collaborative risk management system, parametrized for ports' SC services; the SC partners can analyze all threats arising from their SC cyber assets, including threats associated with port critical information infrastructures interdependencies and associated cascading effects. In the next section, we will outline the main steps of the MITIGATE methodology [31], which is compliant with ISO28000, ISO28001, and ISO27005 and can be used by SCS business partners to estimate the cyber risks of their SCS services at asset level; it lays a special focus on the cascading effects in multisector environments. Special emphasis will be paid in the provision of support for security processes associated with the dynamic (ICT-based) SCs.

MITIGATE RISK ASSESSMENT SCS METHODOLOGY AT ASSET LEVEL

MITIGATE is a supply chain risk assessment (SCRA) methodology that aims to estimate and forecast the cyber risks of any SCS for which its provision/delivery requires the interaction of various cyber assets from different business partners (cross-partners' cyber assets). The objectives of the MITIGATE methodology are these:

- identify and measure all cyber threats within an SC service;
- evaluate the individual, cumulative, and propagated vulnerabilities;
- predict all possible attacks and threats paths and patterns within the SC cyber system (which consists of cross-partners' cyber assets) based upon specific propagation rules that consider the various attackers' profiles described by NIST SP800-30 r1 "Guide for Conducting Risk Assessments";

- estimate the existence of zero-day exploitable vulnerabilities;
- assess the possible impacts;
- derive and prioritize the corresponding cyber risks of the SC cyber assets;
- formulate a proper mitigation strategy.

We made the following assumptions: the business processes of the SCS are linear (not cyclic) and thus the cross-partners' cyber assets are interconnected in one-way, linear paths (thus cyclic attacks are not possible); the cross-partners' cyber assets are used only for the provision or delivery of the SCS and are isolated from the partners' individual ICT infrastructure. We also assume that every SCS is accompanied by a service level agreement (SLA) that includes the standard aspects of the SCS (scope, quality, responsibilities between the service providers, and the service users).

MITIGATE is an SCS risk assessment methodology on asset level. The specific assets and implemented controls of each business partner involved in the SCS are necessary inputs required by the MITIGATE approach. To ensure the privacy and protection of this data, the partners will be asked to provide an Enhanced Security Declaration, which will be a confidential, legally binding document (included in the SCS SLA) that will reveal the commitment of the partners to identify all their organizations' cyber assets that are necessary in the provision of the SCS and the controls they have undertaken. The signed Enhanced Security Declaration will be a prerequisite for the ISO27001, ISO27005, ISO28001, and ISPS (IT section) compliance of the MITIGATE methodology. ISO27001 requires the commitment and legal binding of the organization in formulating an information security management system. ISO27005 requires, in addition, the exact identification of cyber assets and controls. ISPS (in its section "Electricity, communication and telecommunication systems, and computer and network systems") requires the organizations to identify these cyber assets and the controls that have been undertaken.

General Structure of the Methodology

The MITIGATE SCRA methodology is triggered by any specific SCS and follows the standard risk assessment six main blocks, i.e., boundary setting, threat analysis, vulnerability analysis, impact analysis, risk estimation, and mitigation strategy.

The realization of each block contains several steps (Sj) and substeps (Sij) as presented in Table 5.1. The main steps of the proposed methodology are described in detail in the following sections:

Step 0: Scope of the Supply Chain Risk Assessment

The SCS is selected (which triggers the SCRA) as well as the boundaries of the assessment (scope, goal, expected outcome). The risk assessor (which may be any business partner within the SCS) defines the goal, the scope, and the outcome of the assessment.

Step 1: SCS Analysis

In this step, the SCS under examination, as defined in Step 0, will be decomposed. The actors of this step are the risk assessor that initiates the assessment in consensus with the business partners (based upon the SLA and the signed Enhanced Security Declaration Form) executing the following steps:

Step 1.1: Scope and objectives of the SCS: The risk assessor that initiates the risk assessment in consensus with the business partners defines the scope of the SCS under examination and provides the goal, the scope, and the outcome of the SCS.

Step 1.2: SCS business partners and participant's identification: The risk assessor (in consensus with the business partners) identifies the business partners, which participate in the SCS under assessment. In addition, each business partner indicates the participants that will be involved from his/her organization in the current risk assessment.

Step 1.3: SCS modeling: The fundamental goal of this step is the identification and modeling of the main cyber or/and physical (controlled/monitored by a cyber system) processes that comprise the examined SCS. In this context, the risk assessor in consensus with the business partners should perform the following activities:

- SCS's business processes (SCSBPs) identification: All cyber or/and physical processes are defined and recorded.
- Business partners association: The business partners identified in Step 1.2 are linked to the defined SCS business processes.
- SCS cyber assets identification: All cyber assets required for the provision of the examined SCS and the corresponding SCSBPs are identified and reported. The assets can be categorized into various types like application server, desktop application, enterprise/Web application, mobile application, Web server, operating system, Web browser, etc.

Table 5.1 Mapping between SCRA main blocks and substeps

Boundary setting	Threat analysis	Vulnerability analysis	Impact analysis	Risk estimation	Mitigation
S1.1: Goal & Objectives	S2.1: SCS cyber threats' identification	S3.1: Identification of confirmed vulnerabilities	S4.1: Individual asset impact assessment	S5.1: Individual asset risk assessment	S6: Risk mitigation
S1.2: Business partners	S2.2: SCS threat assessment	S3.2: Identification of zero-day vulnerabilities	S4.2: Cumulative impact assessment	S5.2: Commutative risk assessment	
S1.3: Modeling		S3.3: Individual vulnerability assessment	S4.3: Propagated impact assessment	S5.3: Propagated risk assessment	
		S3.4: Cumulative vulnerability assessment			
		S3.5: Propagated vulnerability assessment			

- Assets interdependencies modeling: Specify and illustrate the interconnections that exist between the entities and the assets comprising the SCSBPs by taking into consideration the defined types of dependencies (e.g., hosting, exchange data/information, storing, controlling, processing, accessing, installing, trusted).
- Enhanced business declaration: Identify and report the security controls applied to each asset identified on previous steps.

Step 2: SCS Cyber Threat Analysis

All threats related with the SCS cyber assets reported in the previous step will be identified and evaluated in terms of their likelihood of occurrence. This process includes the following activities:

Step 2.1: Individual Cyber Threats Identification Applicable to the SCS Cyber Asset

In this step, all individual cyber threats against the SCS cyber assets will be identified. The list of SC cyber threats will be completed as follows: All SC business partners (from their expertise) will identify specific cyber threats that are applicable and relevant to the SCS cyber assets comprising the SCS under examination. Furthermore, from social media, additional cyber threats may be found and added. Existing repositories of cyber threats will offer additional threats to be added. Crowdsourcing (the process of harnessing/seeking the contributions of a community of online users/security experts/stakeholders) will aid in the identification of cyber threats.

Step 2.2: SCS Threat Assessment

In this step, the probability of occurrence of each cyber threat to each cyber asset is assessed. The methodology uses a five-tier scale (Very Low = 20%, Low = 40%, Medium = 60%, High = 80%, Very High = 100%), and the evaluation is based upon the following criteria: (1) the expected frequency of appearance based on the history of previous incidents; (2) the participants' intuition and knowledge; and (3) information retrieved from social media and existing repositories that will be used to gather peripheral information and draw conclusions. Assume that for the threats T, a participant believes, based upon his expertise, that the probability of occurrence of the scenario T to asset A is about 50% (corresponding to the threat level "Medium"). Taking into consideration the information retrieved from social media, he concludes the probability is about 70%. In addition, he replies that T has a history of one incident in the last 12 months (corresponding to the threat level "High"). Thus, based on the answers of the participant and on the used scales, the threat level assigned is High = 80%.

Step 3: Vulnerability Analysis

The purpose of a vulnerability analysis is to identify, quantify, and prioritize the vulnerabilities that exist in the cyber assets of the SCS under examination. Vulnerabilities could be induced through poor configuration, lack of security patching, etc. The analysis consists of the following steps:

Step 3.1: Identification of Confirmed Individual Vulnerabilities

Confirmed vulnerabilities (vulnerabilities that have not been treated by the business partners) corresponding to each asset will be found as follows: from the online databases (e.g., https://web.nvd.nist.gov, http://www.cvedetails.com); from open source and commercial vulnerability scanners (e.g., OpenVas); and from the list of the applied security controls identified in the enhanced business partner declaration (see Step 1.3).

All identified confirmed vulnerabilities will be accompanied with specific attributes (access vector, access complexity, authentication, threat and vulnerability categories, exploitable) defined by the common vulnerability scoring system (CVSS).

Step 3.2: Identification of Potential/Unknown (Zero-Day) Vulnerability

In this step, we predict additional vulnerabilities; we seek to investigate the probability of existence of zero-day vulnerabilities exploitable in the assets. Given the matter-of-fact limitations of vulnerability repositories, viz. they cater for a subset of existing software applications, there are delays in the confirmation and publication of a vulnerability discovery, commercial vulnerability repositories trickle down confirmed vulnerabilities to open or free repositories to maintain their added value, etc., it is important to model for unknown and/or undisclosed vulnerabilities to gain a pragmatic view of the SC's risk exposure.

The estimation can be made over all time scales in the available dataset by empirically characterizing the distribution of the vulnerabilities lifespans and determining the number of vulnerabilities publicly announced for a specific period (e.g., using the rate of vulnerability announcements in the National Vulnerability Database). Examples include the following:

- Estimate how many (e.g., zero-day) vulnerabilities will be disclosed or may exist in different assets categories (e.g., operating systems).
- Estimate how many (e.g., zero-day) vulnerabilities will be disclosed for each threat/vulnerabilities category.
- Crowdsourcing and social media can also provide to the assessor information about potential (i.e., not confirmed) vulnerabilities. Zero-day vulnerabilities are modeled in the same way as confirmed vulnerabilities.

Step 3.3: Individual Vulnerability Assessment

This step aims at estimating the severity of all (confirmed and zero-day) vulnerabilities identified in the previous steps (Step 3.1 and Step 3.2). In this way, we calculate the probability of successfully exploiting each vulnerability when all the prerequisite conditions required are met. In MITIGATE methodology, we rely on qualitative values for representing the individual vulnerability level (IVL) and use a five-tier scale ranging from "Very Low" (VL) to "Very High" (VH).

For the calculation of the vulnerability levels we use the CVSS metrics (the access vector (AV), the access complexity (AC), and the authentication (Auth)) retrieved from the online databases; the mapping from the CVSS metrics onto one category of the five-tier is presented in Table 5.2.

For example, taking a vulnerability V with an AV "Adjacent," an AC "Medium," and an Authentication "Single," the resulting IVL coming from the mapping is "Medium."

Step 3.4: Cumulative Vulnerability Assessment

The key limitation of the individual vulnerabilities levels (produced in Step 3.3) is that they can depict the probability that the corresponding vulnerabilities will be exploited successfully when exposed to an attacker without considering the individual actions that the attacker should perform to satisfy the preconditions required for their exploitation.

Thus, the goal of this step is to accurately reflect the exploitation level of the confirmed and zero-day vulnerabilities (identified in Step 3.1 and Step 3.2, respectively), by taking into consideration the IVLs (produced in Step 3.3), the context within which these vulnerabilities appear (e.g., the assets interdependencies/interconnections), and the attacker's profiles.

Thus, this step measures the probability that an attacker can successfully reach and exploit each of the (confirmed and zero-day) vulnerabilities identified in the previous steps in each vulnerability chain (the chain of sequential vulnerabilities on different assets that arise from consequential multistep attacks).

In this context, the following activities should be performed:

- *Entry points identification*: The business partners must define the entry points (assets from which the attacks will be initiated; these assets are considered more reachable by an attacker).

- *Target points identification*: The business partners must define the target points (the assets that are considered a target for attacks due to their criticality).
- *Identify attacker profile*: Attacker profiles will be identified by their relationship with the organization (insider, outsider), their skills (ICT skilled, premature), and their target, e.g., each attacker will be characterized by the pair (insider, premature, target = specific asset), (insider, ICT skilled, target = no target, general damage).
- *Generate vulnerability chains*: This step follows a rule-based reasoning approach (filters) to generate vulnerability chains initiated from the entry points to exploit the vulnerabilities of the target points. We use a set of rules to decide whether an asset can be compromised to be used as an intermediate step to induce further attacks. To this end, one can deduce the following:
    - the accessibility of a vulnerability, we use two metrics: (1) the vulnerability AV's attribute that shows how a vulnerability can be exploited, and (2) the dependency AV's attribute that shows how the assets are connected;
    - For the complexity degree of a vulnerability, we consider the attackers' profiles, i.e., the skills and characteristics the user grants the attacker and the vulnerability AC attribute;
    - To decide whether an asset can be used as a stepping stone to launch further attacks on other assets, we consider the technical effects and impact of the vulnerability;
    - For the exploitability of a vulnerability, we take into consideration whether there is a known exploitation technique for the vulnerability.
- *Select most accurate vulnerability chains*: The business partners review the produced chains and select the most doable (deleting the ones that are not probable or reasonable) chains.
- *Measure cumulative metrics*: Calculate the various types of vulnerability levels:
    - **Individual chain vulnerability level (ICVL)** measures the probability that a vulnerability z that resides in a target point can be exploited given the specific vulnerability chain originated from an entry point with vulnerability v. The calculation of the ICVL is based on the attacker's capability (very low, low, moderate, high, and very high) together with the IVL of an asset/vulnerability combination. In this context, we use a table-based mapping (see Table 5.3) to estimate the

Table 5.2 Mapping of the CVSS metrics on the MITIGATE vulnerability level

AV	Local			Adjacent			Network		
AC auth	**High**	**Medium**	**Low**	**High**	**Medium**	**Low**	**High**	**Medium**	**Low**
Multiple	VL	VL	L	L	M	M	M	H	H
Single	VL	L	L	M	M	M	H	H	VH
None	L	L	M	M	M	H	H	VH	VH

Table 5.3 Mapping of the attacker's capability and the IVL onto the likelihood of exploitation

Capability					
IVL	**Very low**	**Low**	**Moderate**	**High**	**Very high**
Very low	VL	VL	L	L	M
Low	VL	L	L	M	H
Moderate	L	L	M	H	H
High	L	M	H	H	VH
Very high	M	H	H	VH	VH

likelihood that an attack can exploit a specific vulnerability. For this estimation, we use the five-tier scale ranging from "Very Low" (VL) to "Very High" (VH). Therefore, this would mean that if the adversary's capabilities are "Moderate" and the IVL of a vulnerability is specified as "High," then the attacker will succeed in exploiting that vulnerability with "High" likelihood.
In this context, the derived qualitative values are mapped onto probabilities using the mapping given in Table 5.4. Therein, each category is assigned a representative value as well as a range, consisting of an upper and a lower limit. Thus, the category "Very High" is represented by the value 0.93, and the category "High" is represented by the number 0.75. These representatives can be used for the computation of the ICVL by multiplying all values of the vulnerabilities involved in the chain. Afterward, the result can be mapped back onto one category using the predefined ranges. For example, the ICVL is "High," if the computed value lies between 0.65 and 0.84.

- The **entry's chain vulnerability level (ECVL)** measures the probability that a vulnerability z that resides in a target point can be exploited given all independent vulnerability chains originated from a specific entry point considering all its vulnerabilities v. To compute the ECVL, all ICVLs from the entry point V of asset A to the target point are taken into consideration. To this end, the maximum approach is applied, considering the worst case and taking the highest level of all ICVLs.

Table 5.4 Description of the probability scale in MITIGATE

	Representative	
Qualitative values	**Range**	**Number**
Very high	0.85–1.00	0.93
High	0.65–0.84	0.75
Moderate	0.35–0.64	0.50
Low	0.15–0.34	0.25
Very low	0.00–0.14	0.07

- The **cumulative vulnerability level (CVL)** measures the probability that a vulnerability z that resides in a target point can be exploited given that all independent vulnerability chains originated from all entry points.
  For computing the CVL, we analyze all attack paths from all potential entry points. Thus, as for the ECVL, the maximum approach also used here takes the highest level of all ECVLs.

Step 3.5: Propagated Vulnerability Assessment

This step measures the conditional probability that an attacker can successfully reach and exploit each of the individual vulnerabilities (identified in the previous steps) in each vulnerability chains. The following activities should be performed:

- **Entry points identification**: The business partners must define the entry points, i.e., the assets from which the attacks will be initiated and are considered more reachable by an attacker.
- **Chain length** (l) definition: The business partners must define the chain length l according to the original SCS model (see Step 1.3).
- **Identify attacker profile**: Attacker profiles will be identified by their relationship to the organization (insider, outsider), their skills (ICT skilled, premature), and their target, e.g., each attacker will be characterized by the pair (insider, premature, target = specific asset), (outsider, premature, target = no target, general damage).
- **Generate propagated vulnerability chains**: This step follows a rule-based reasoning approach (filters) to generate consequential chains of vulnerabilities on different assets (raised by multistep attacks initiated from the entry points). These chains are generated similarly to the vulnerability chains mentioned in Step 3.3.
- Select the most accurate **propagated vulnerability chains**: The business partners review the produced chains and select the most appropriate chains.
- **Measure propagated metrics**: The following propagated metrics are calculated:
  - *Individual propagated vulnerability level (IPVL)* measures the weakness of an individual propagated vulnerability chain to be exploited due to exposure of a specific vulnerability v in an entry point.

- *Propagated vulnerability level (PVL)* measures the weakness of all independent propagated vulnerability chains to be exploited due to exposure of a specific vulnerability v in an entry point.

  It should be noted that whereas for the CVL a specific target point is specified, the PVL requires a specific entry point where the attacker is supposed to start from and an attack path length the user wants to consider in his examination. To this end, the calculations of the IPVL and PVL have been performed similarly to the calculation of the ECVL and CVL, respectively.

Step 4: Impact Analysis

Impact analysis refers to the assessment of the expected consequence if a malicious action is realized. This analysis includes the following steps:

Step 4.1: Individual Asset Impact Assessment

This step estimates the impact of the successful exploitation of each individual confirmed vulnerability (identified in Step 3.1) and zero-day vulnerability (identified in Step 3.2) to an asset. In this context, for each vulnerability on the asset the CVSS metrics retrieved from the online databases are used to calculate the impact level. The impact level measures the effect that can be expected because of the successful exploitation of a vulnerability v that resides in an asset.

Similarly to the IVL in Step 3.3, we define a mapping from the three security criteria, confidentiality (C), integrity (I), and availability (A), covered in the CVSS impact metric onto the five-tier scale ranging from "Very Low" (VL) to "Very High" (VH) (see Table 5.5). This provides a single estimation for the overall impact of a specific asset/vulnerability combination.

For example, taking a vulnerability V with an impact "Partial" on the confidentiality, "Complete" on the integrity, and "None" on the availability, the resulting impact level coming from the mapping is "Medium."

Step 4.2: Cumulative Impact Assessment

According to the different vulnerability levels used in the methodology (i.e., the IVL, the ICVL, the ECVL, the CVL, and the PVL), there are also several different impact levels defined.

The cumulative impact level is defined as the impact that occurs after a specific asset/vulnerability combination has been exploited by an attacker using any possible entry point (cf. the CVL in Step 3.4).

Table 5.5 Mapping of the CVSS metrics on the MITIGATE impact level

C	None			Partial			Complete		
IA	**None**	**Partial**	**Complete**	**None**	**Partial**	**Complete**	**None**	**Partial**	**Complete**
None	VL	VL	L	L	M	M	M	H	H
Partial	VL	L	L	M	M	M	H	H	VH
Complete	L	L	M	M	M	H	H	VH	VH

Table 5.6 Mapping of the impact level and the ICVL onto the individual chain impact level

Impact					
Level ICVL	**Very low**	**Low**	**Moderate**	**High**	**Very high**
Very low	VL	VL	L	L	M
Low	VL	L	L	M	H
Moderate	L	L	M	H	H
High	L	M	H	H	VH
Very high	M	H	H	VH	VH

In this context, taking into consideration the impact levels calculated in the previous step, we derive the following values:

- The **individual chain impact level (ICIL)** reflects upon the effects of the successful exploitation of the vulnerability z of a target point given that a specific vulnerability chain that originated from a specific entry point with vulnerability v has been exploited. For the computation of the ICIL, the individual impact of the target point and the ICVLs for this specific target point are required. These serve as input to the mapping given in Table 5.6. Thus, if the impact level of an asset/vulnerability is "High" and the ICVL to reach and exploit the respective vulnerability is specified as "Low," then the ICIL of that vulnerability is classified as "Moderate."
- The **entry's chain impact level (ECIL)** reflects upon the effects of the successful exploitation of the vulnerability z of a target point given that all independent entry chains starting from the vulnerabilities v of a specific entry point have been exploited.
- The **cumulative impact level (CIL)** reflects upon the effects of the successful exploitation of a vulnerability z of a target point given that all independent entry chains starting from all entry points have been exploited.

  As for the ECILs and CILs, the maximum approach is applied, considering the worst case and taking the highest level of all ICILs and ECILs, respectively.

Step 4.3: Propagated Impact Assessment

The propagated impact level (PIL) is defined as the overall impact that occurs when an attacker exploits a specific asset/vulnerability combination and further moves on into the network starting from this entry point (cf. the propagated vulnerability assessment in Step 3.5). In other words, this relates to the damage an attack can cause at any asset/vulnerability combination on its way through the network. In this step, the following impact levels are calculated:

- The individual propagated chain impact (IPCI) reflects the effects of the exploitation of the vulnerability v that resides in the entry asset given that a vulnerability v may cause the exploitation of a vulnerability chain.
- The PIL reflects the effects of the exploitation of the vulnerability v that resides in an asset given that a vulnerability v may cause the exploitation of a set of independent vulnerability chains.

  The IPCIs and PCIs are computed similarly to the ICILs and CILs described in Step 4.2.

Step 5: Risk Assessment

In this step, various risks estimations will be performed as follows.

Step 5.1: Individual Asset Risk Assessment

After collecting all impact values (Step 4.1), threat levels (Step 2.2), and vulnerability levels (Step 3) for a specific asset, we calculate the following:

- The individual risk level represents how dangerous a threat s is to a specific asset within an SCS.
- The individual risk metric represents how dangerous all threats are to a specific asset.

Step 5.2: Commutative Risk Assessment

The goal of this step is to accurately reflect the risks that the commutative vulnerability chains introduce to the SCS. In this context, the following metrics are calculated:

- The individual commutative risk metric refers to the risk exposure of a threat s occurring due to a vulnerability z that exists in a target point given that a specific vulnerability chain originated from an entry point with vulnerability v has been exploited.
- The entry's chain risk level refers to the risk of a threat s occurring due to a vulnerability z that exists in a target point given that all independent entry chains starting from the vulnerabilities v of a specific entry point have been exploited.
- The commutative risk level refers to the risk of a threat s occurring due to a vulnerability z that exists in the target point given that all independent entry chains starting from all entry points have been exploited.

Step 5.3: Propagated Risk Assessment

The goal of this step is to accurately reflect the risks that the propagated vulnerability chains introduce to the SCS.

- The individual propagated risk level refers to the risk of a threat s occurring due to a vulnerability v that exists in an entry point given that v may cause the exploitation of a specific vulnerability chain.
- The propagated risk level refers to the risk of a threat s occurring due to a vulnerability v that exists in an entry point given that v may cause the exploitation of all independent vulnerability chains.

Step 6: Risk Mitigation: Selection of Security Controls

In this step, the risk assessment values are compared against specific criteria (set and agreed to by all business partners) to select additional security controls required by the business partners and by the SCS (as, a whole) meeting the thresholds that have agreed upon in their Enhanced Security Declaration. MITIGATE methodology was implemented in the MITIGATE system, a specific set of services deriving from diverse software artefacts that have been tightly integrated in a harmonized system. Prior to the harmonization process, a concrete componentization scheme has been defined in the context (as described in the next figure). As depicted in Fig. 5.1, there are seven main components that comprise the MITIGATE system, namely (1) the Asset Modeling and Visualization, (2) the Supply Chain Service Modeling, (3) the Simulation and Game Theory, (4) the Collaborative Risk Assessment, (5) the Open Intelligence and Big-Data Analytics, (6) the Notification and Reporting, (7) the Administration, and (8) the Access Control and Privacy Component, which is not depicted on the next figure since it is totally horizontal to all the other components.

These components are logical since they represent a grouping of functional requirements that coordinate the process of the platform development. A snapshot of the MITIGATE dashboard can be viewed in Fig. 5.2. The delivered functionality is organized through a menu that does not correlate one to one with the components of the preceding figure since a conceptual grouping has been performed, for the sake of a clear value proposition. More specifically, the main MITIGATE menu bar is organized in the following groups:

- the group of Open Intelligence where the online synchronization of sources is controlled,
- the group of Threat, Control, and Vulnerability management,
- the group of the Asset management.

The fact that the final platform is unified raised many issues during the development process since the existing components rely on diverse technologies. It is self-explanatory that a harmonized platform can be created only through the adoption of a common technological base. Spring was used to build the platform, which is a state-of-the-art Java enterprise framework. Although Spring can support various patterns of development the Inversion of Control pattern has prevailed. As a result, the final MITIGATE platform relies on Java 8 as the base programming language and on Spring as the enterprise framework.

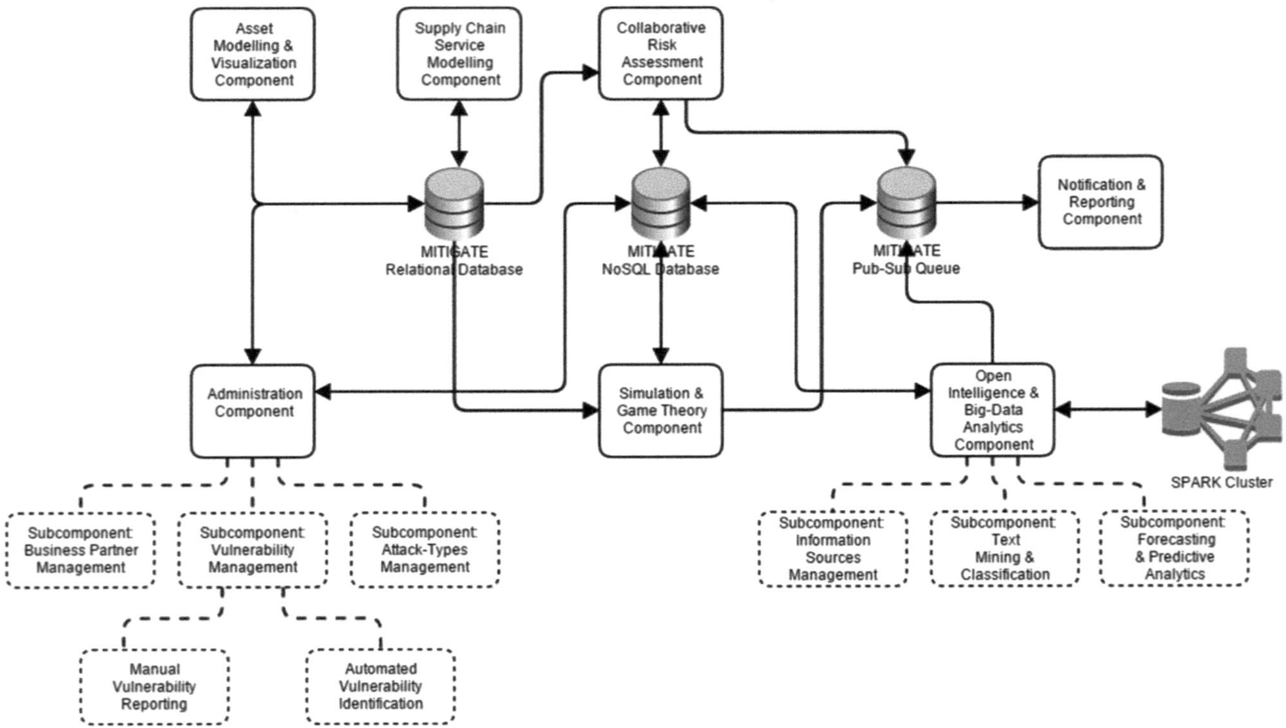

Figure 5.1 MITIGATE high-level architecture.

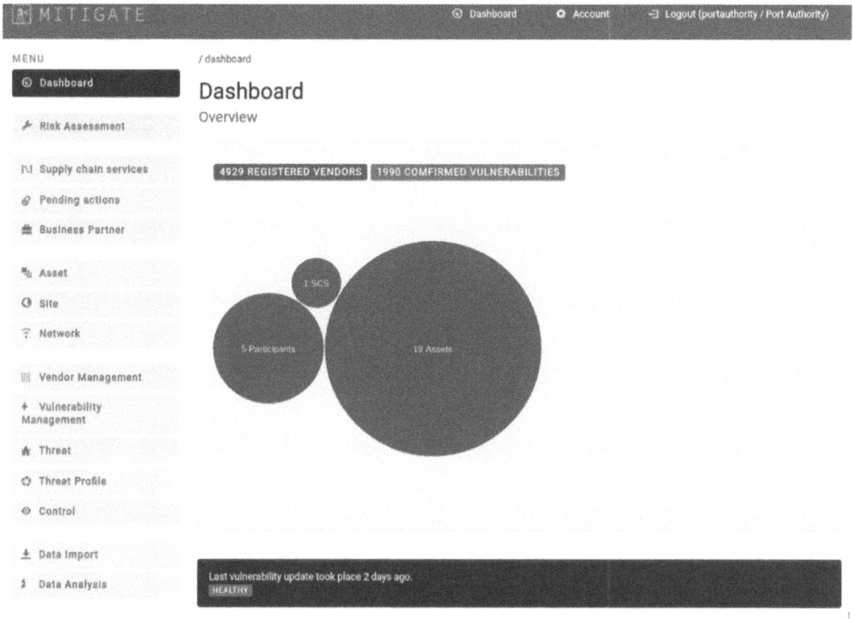

Figure 5.2 MITIGATE dashboard overview.

It relies on a completely separate UI and business logic tier and the Thymeleaf [3,83,93,95,153] templating engine. The UI consists of native HTML5 that is rendered through the Thymeleaf engine.

For the persistency layer the Neo4J6 engine has been used, which is classified as a graph database. The attack chains represent the different paths that a hacker can follow to harm a specific asset. Practically each path consists of nodes that can be exploited using a specific vulnerability of the node. Since one node may entail multiple vulnerabilities the generation of the chains and the processing of the chains during the risk calculation is extremely computationally intensive. Neo4J is built dedicated for graph scalability.

Finally, most of the computations that have to be performed for the open intelligence component require the usage of a big-data computation framework. Among many frameworks, the Spark has been selected as it is considered one of the most prominent ones. Furthermore, it is compliant with Java 8. An extremely useful feature of Spark is its extensive library for language processing. This library has been thoroughly used in the context of open intelligence during the classification of cyber sources.

The MITIGATE system adopts the industry standards (CVE and CAPEC) for vulnerability and exposure names to support well-recognized vulnerability identifiers. The benefits of having the CVE and CAPEC identifiers are significant in terms of information timeliness and data accuracy and reliability.

CHAPTER 6

Conclusions and the Way Forward

Commercial ports are characterized by a dual "cyber"/"physical" nature: their physical attributes are related with the port infrastructure, including facilities, buildings, platforms, gates, marinas, and data centers, while their cyber attributes are related to their ICT systems (e.g., networks, ICT hardware equipment, port community systems [PCS], services, data, users, procedures). Existing maritime security standards concentrate on the protection of the physical nature (safety) of the ports ignoring their cyber nature (security), leaving the ports unprotected.

We have witnessed the emergence of initiatives and research efforts that attempt to deal with the risks and vulnerabilities of the ports' critical information infrastructures (CII), addressing their dual nature, the complexity and interdependencies of their physical and cyber assets. This started with the first ENISA (European Union Agency for Network and Information Security) report on cyber maritime security (2011) in which the low cybersecurity awareness level in the maritime sector was reported. Various projects (e.g., CYSM, MEDUSA, MITIGATE, 2013–18) since then investigated various cybersecurity issues in the maritime sector. From this experience, the author reported in this book the main findings that lead to various open issues and challenges (technical, policy, legal, business) drawn in this conclusive chapter.

Commercial ports host, store, or exchange information and data to provide maritime participants with electronic and mobile port and maritime services. It is obvious that many problems that occur in ports are associated either with lack of information or absence of automation and secure integration of unified maritime operations of the provided services.

Ports and supply chain providers need to improve their security culture by realizing the value of information security to support their ports' operations, supply chain services (SCSs), their business models, and business collaborations. Concrete, measurable training and awareness actions need to be undertaken at all levels (national, regional, European, international) to train all participants in maritime logistics and supply chains on cybersecurity, raising their awareness and to build a security culture in the provision of secure maritime operations and services. Maritime organizations (e.g., IMO, NATO, EMSA, ENISA, IPCSA, DGMARE) may contribute toward a rise

Port Cybersecurity
ISBN 978-0-12-811818-4
http://dx.doi.org/10.1016/B978-0-12-811818-4.00006-X

in security awareness by giving presentations during various events and offering security seminars, certification courses, and summer schools. Furthermore, they may use their dissemination tools (e.g., newsletters, magazines, newspapers, social media) to inform the maritime world about emerging security threats.

Insurance companies and auditors may be the main drivers in the campaign for maritime security awareness by enforcing maritime providers to comply with existing cyber and supply chain security standards (e.g., ISO28000, ISO28001, ISO27001, ISO27005), directives (e.g., NIS, eIDAS, General Data Protection Regulation [GDPR]) and cybersecurity guidelines (e.g., IMO guidelines on cybersecurity).

The coordinated implementation of the EU Directive 2010/65/EU23 by establishing guidelines on the "national single windows" needs to include security risk assessment to avoid security incidents with great impact at national, EU, and international levels. The United States, 2016H.R.3878, the European Cybersecurity Strategy, and NIS directive embrace the ports CII to establish an open, safe, and secure cyberspace. They highly contribute to coordinated prevention, detection, and mitigation of risks enabling mutual assistance and information sharing among them, the maritime authorities, and the national CERTS/CSIRTS.

With the EU GDPR, which will come into effect in May 2018, Europe has made an important step toward strengthening and clarifying the legal framework that protects the privacy of its citizens. One of the key concepts in this regulation is "privacy by design," which implies that privacy should be involved from the start in building maritime systems and developing port services. However, at this stage, it remains a challenge to provide technical specs on how to implement this concept in the maritime systems and software and furthermore how to engineer and audit these systems that offer privacy by design.

The complex ports' CIIs and their SCSs (e.g., container management, vehicle transport, LNG, cruise reservations) involve various business partners, e.g., ministries, maritime companies, ship industries, customs agencies, maritime/insurance companies, and other CII (railways, airports, energy networks, and telecommunication). The plethora of complex interdependencies within these maritime business entities results in threats with catastrophic impacts beyond the provision of the SCS and the business partners involved; the whole maritime ecosystem, the economic environment, and the safety of people and involved nations may be affected. Innovative dynamic modeling techniques are needed to capture the interdependent business processes, assets (physical and cyber), threats, and their cascading effects involved within the ports' CIIs and their maritime business partners. Although the 2016 IMO cybersecurity guidelines for maritime cybersecurity management provide high-level recommendations to safeguard shipping from current and emerging cyber-related threats and vulnerabilities, they do not provide guidelines for the port authorities and the supply chains to manage the security of their physical/cyber interconnected assets.

Although there is a plethora of cybersecurity standards (e.g., ISO27001, ISO27005, NIST Framework) and SCS standards and conventions (e.g., ISO28000, ISO28001, ISPS), there are no specific user-friendly ICT tools combining and implementing these standards that the ports or any business partner in the maritime logistic chain can use to estimate and manage the risks and their cascading effects. The Supply Chain Security Standard ISO 28001 provides generic guidelines as an informative reference for organizations seeking to implement or refine a specific methodology. It does not directly address ports' SCS needs and specificities. The EC research projects CYSM, MEDUSA, and MITIGATE provide enhanced risk assessment methodologies and tools targeting the maritime sector. Although they need large-scale testing and commercialization, they can serve as best practice examples for the maritime security industry.

The biggest problem is the security of the industrial control systems (e.g., SCADA system), commonly used in the maritime world. There are not available targeted risk assessment tools or open repositories with SCADA vulnerabilities and controls to be able to estimate the actual risks of these systems.

All existing maritime security projects are limited to risk assessment functionalities and do not offer security incident management handling services. Neither the CERTs nor the related agencies (e.g., European Maritime Security Agency [EMSA]) and organizations (e.g., IMO) offer to the port operators and maritime logistics and supply chain providers dedicated security incident management handling services. These limitations put port and supply chain operators under pressure to report existing attacks, to access/receive warnings for upcoming attacks and vulnerabilities, and receive technical assistance and guidance on investigating and handling security incidents. There is an urgent need for the formulation of security incident management handling centers to help the maritime sector to do the following: (1) thoroughly assess the vulnerabilities of all maritime interconnected physical and cyber assets; (2) continuously forecast and evaluate the probability of hybrid (physical, cyber, and combined) attacks; (3) access/

receive warnings for upcoming attacks and vulnerabilities; (4) easily recreate, visualize, and forecast propagation and cascading effects of attacks in their maritime services and operations and anticipate how these attacks propagate across all interconnected maritime CI and business infrastructures; (5) follow a targeted step-by-step framework providing timely technical assistance and guidance on investigating and handling complex, interrelated physical/cyber maritime security incidents; (6) follow targeted forensics methodologies and pointers for appropriate open source forensics tools that will guide them to deeply identify and investigate a security incident and extract relevant evidences and information; (7) combine and analyze all security incident-related information and proofs in an effective and accurate manner; and (8) receive guidelines on how to respond to a security incident and share information and warnings with the public in the vicinity of the maritime installations.

PCS are complex electronic platforms that connect multiple systems operated by a variety of organizations in the maritime ecosystem and a plethora of ICT providers; PCS security is treated as a "black box." National single windows, which are based upon the PCS, need to be assessed, and accompanied security policies should be audited to guarantee their trustworthy operation. Unfortunately, most of the port operators and other SCS providers use nonstandardized, nonharmonized security management practices concentrating on physical threats and weaknesses, ignoring the cyber ones. Security incidents (both physical and cyber) might be flawed or counterfeit, thereby jeopardizing the operation of the ports and the whole supply chain, causing economic disruption and disruption to transport systems and to international trade. Not only PCS but all maritime software is always treated as a "black box"; maritime operators and facilitators do not question the security of their infrastructures, although they are not aware of their security mechanisms. They consider it a secure black box. No security audits are performed by certified national, European, or international authorities. As a result, no one except the manufacturers know the security level and vulnerabilities of the maritime software. Code auditing needs to be performed by certified authorities to the maritime software so they may be accompanied by respected security certificates.

Automated border control (ABC) systems meet the need to increase throughput at border control points while maintaining the requisite levels of security. ABCs rely on biometrics (face, eyes, fingerprints) to verify the identity of the travelers. ICT technologies then authenticate the identity real time by matching the characteristics of individuals against previously stored records. ICAO recommends facial recognition as the globally interoperable biometric technology for machine-assisted identity confirmation, while acknowledging that some authorities may supplement this with fingerprint and iris recognition. e-Passports contain traveler data (including the biometric markers) inside an embedded chip. This chip has been designed with different data protection mechanisms in place to ensure that only authorized parties can access the information contained inside.

The FRONTEX requirements (FRONTEX, 2012) for the ABC systems and the requirements imposed by the Practical Handbook for Border Guards (Schengen Handbook), 2006 need to be enhanced to address the remaining challenges:

- mobility: capable of being installed in various locations of the port (e.g., on ship, arrival/departure areas, on vessels);
- interoperability: adequate interoperable number of ABC gates to ensure that checks are carried out quickly;
- multimodality–interconnection: different nature of interconnected ABC gates need to be installed in the ports; indoor ABC gates in the ports' indoor terminals (if exist) for passengers and crew; mobile ABC gates on board ship or in an area set aside; ABC portable devices held by the patrol controllers in case the ship arrives outside the port terminal and the border guard needs to reach it by boat; large ABC gates that vehicles, heavy goods vehicle drivers, and any accompanying persons need to be checked; ABC systems are optimally used in this case in the entry/exit points for trucks and containers.
- security and privacy: ABCs become more and more susceptible to current threats; they are prone to snooping, interruption, vandalism, and interception of the ABCs. The ABCs need to be compliant with all relevant security directives, need to have their own security and privacy policies and procedures, and follow the ISO27001, ISO27005 standards (since they are ICT assets).

Accountability needs to accompany security, and accountability measures need to be implemented in the maritime software and ICT systems. Certification efforts in maritime cybersecurity products need to take in the security, privacy, and accountability levels.

Security needs to be treated as a shared responsibility among all maritime providers. Service level agreements need to be enhanced to include security responsibilities, business continuity, and disaster recovery responsibilities. No single maritime entity can afford the financial burden to manage (prevent, detect, analyze mitigate) security incidents in their organization;

their providers need to share the security responsibility. For example, if a navigation system is vulnerable to snooping, interruption, or interception, the provider of this system needs to become responsible for the mitigation of its vulnerabilities and not the maritime company or the port authority.

Economic crisis has brought privatization to EU ports, and foreign investors manage the EU port systems under their own security practices. This is another threat for business espionage having negative impact to the EU economy. The EU needs to enforce security practices to EU ports. EU auditing authorities and the insurance companies should audit the security policies of the ports and their compliance with the EU directives and international standards.

Security challenges are not limited to the maritime borders of a single country or region. The aim should be a synergetic international response, with harmonized implementation of the existing EU and international rules (e.g., ISPS, EC Regulations No 725/2004, 2005/65, 2013/0027, 2015 European Parliament Proposed Directive DATAPROTECT 235, and NIS/2015). The transnational cooperation for enhancing ports' security has been emphasized in the Regional 2014 European Union Strategy (EUSAIR). Furthermore, IMO has NATO's 2015 annual reports underline the importance in building global synergies for emerging threats, security incidents, and hazards.

Public key infrastructures (PKIs) have been qualified as an appropriate framework for the provision of all security dimensions (data confidentiality, integrity, availability, and nonrepudiation) in several business sectors. These measures are offered through relevant PKI-based security mechanisms (e.g., data encryption, digital signature, time-stamping), which can assure secure electronic transactions. PKI serves as solution to the authentication of IoT devices (e.g., sensors) currently used in the maritime business (e.g., container management).

One of the main current PKI needs is interoperability, which makes possible the secure interconnection and cooperation between different PKI structures, thus enhancing their feasibility and applicability at regional, national, as well as international level. This is even more important in the maritime sector, due to the increased demand for mobility of maritime participants (physical persons or "things") and the criticality of the port data, which make the secure and interoperable information exchange a basic element for high-level port service provision.

PKI interoperability is usually addressed through the cross-certification service, which can be described as the way to establish chains of trust between different certification authorities (CAs). However, CAs have not been yet established for the maritime sector, and cross-certification is not yet technically provided in an automated way, resulting often in a difficult and time-consuming paper-based process, which reduces the flexibility and usability of the method itself.

The lack of automated cross-certification is at a great extent due to the inadequate standardization of the certificate policies (CPs), which define the PKI certificates' profile and thus form the basic comparison criteria for the mutual acceptance of national CAs. More specifically, although the CP structure is defined in some of the existing standards, there is still a significant gap in the standardization of the CP content (e.g., roles of the involved subjects, certification, and registration requirements). In addition, there is no systemized way for the development and the comparison of CPs, making their comparative analysis a difficult task. The mentioned restrictions, together with the lack of the necessary legal/regulatory PKI harmonization, disable the automated CP comparison, obstructing in this way the automation of the overall cross-certification service and, thus, making difficult the secure electronic cooperation, information exchange, and knowledge sharing.

Another major problem is that the technologies used in the utilization of e-port services are not assessed from a security aspect, and security measures are not implemented. For example, radio frequency identification (RFID) technologies are commonly used in port logistics, container identification, and cargo management. Several leading port facilities have already become, or have contracted to become, RFID enabled. However, RFIDs face numerous threats, e.g., eavesdropping on RFID readers is a major threat [139]. It can be a highly effective form of corporate espionage. RFID readers themselves can broadcast RFID tag data over long distances, often up to hundreds of meters away. This means that an eavesdropper can gather the RFID tag information and can use it or modify it.

Maritime activities were severely affected by the most recent economic crisis, but at the same time, they continue to be at the core of the regional and international economic growth. Cross-border, efficient, trustworthy e-port services will accelerate their presence in the digital maritime market improving the economy of the sector and the global maritime markets. If ports harmonize their digital security practices ensuring information sharing among them and the appropriate CERTS/CSIRTS, they will create the necessary intelligence to better address security incidents and manage existing and upcoming threats. However, the ports do not have the resources to invest and enhance their e-port services and become antagonistic.

The adoption of cloud computing technology can bring a new revolution to the maritime field enabling a holistic collaborative environment offering cross-border, trustworthy e-services to all commercial ports and their users. The main advantages that the ports will merit by adopting a cloud computing approach will include a reduction of development and operational costs; scalability of the services to be used by many users; reduction of execution and response time; quality of service; security, privacy, and interoperability issues will be resolved centrally since they will be handled by the cloud provider; easy access via a browser or plug-in, multimodal access (mobile terminals, PCs, etc.) from anywhere, anytime; easy to use since no specialized knowledge will be required for using services or applications; and economy of scale by saving time, cost, and human resources.

There are various types of clouds that can be implemented in the maritime sector: a *local port cloud* limited within the boundaries of a port authority; a maritime *private cloud* can be implemented at local level offering e-port services to local ports; a *regional port cloud,* which will be a private cloud used by other regional/national ports of similar interests; and a *maritime community cloud* can be implemented at a continent/international level offering cross-border e-port services to international ports.

The implementation of a cloud infrastructure may be cost effective since there are many open source solutions available, e.g., OpenStack [9], Eucalyptus [10], OpenNebula [15], Nimbus [43], Ubuntu [78], EyeOS [79], Collectd [19], and BitNami Cloud Hosting [97] may be considered by a maritime cloud provider to build a cost-effective cloud infrastructure.

The security and privacy management, monitoring, and auditing could be provided to all commercial ports by a maritime cloud service (*security as a service*) offered by a trusted maritime organization (e.g., IMO, NATO, EMSA). This will enable the provision of security and privacy based on uniform standards and methodologies, enabling cross-certification and provision of trustworthy, cross-border maritime services.

Identity management can also be provided as a cloud service, adopted by a maritime central authority enabling the authentication of the maritime participants and the issuance of authorization credentials and tokens. In this centralized option, a maritime authority, acting as cloud provider, can undertake the management of different namespaces, authentication of all participating entities, and issuance of authorization tokens using its central identity management system.

Another option is the *federated* alternative where local, national, and regional maritime cloud providers agree to trust the authentication and authorization information that they share and establish rules and policies to make sure that this trust can be managed. This collaboration of trust is known as federation. There are numerous technological solutions for implementing centralized and federated identity management that can be used by potential maritime cloud providers. These technologies define a set of protocols for the secure exchange of identity information that can be used between cloud providers. Among the common standards that they can be based upon are SAML [114] and Liberty Alliance [118], WS-Federation [117], and WS-Trust [157]. Maritime cloud services would solve the interoperability problem of cross-border services as well enhancing the maritime market sector in a holistic approach, contributing toward the implementation of an integrated maritime policy.

To conclude, better worldwide maritime cybersecurity governance is needed, i.e., a sustainable, operational cybersecurity maritime governance center that thru its collaboration with all existing cybersecurity stakeholders (agencies, centers etc.) will provide worldwide acknowledged governance by promoting the following areas:

Maritime cyber education and cyber skills development: Provide appropriate input to marine education policymakers to enhance their education strategies and implementation plans (e.g., more rigorous cybersecurity curricula) for building cybersecurity awareness and education improving the quality, consistency, relevance to maritime industrial, research, and societal needs. Continuously assess the cybersecurity professional certification programs.

Awareness rising: Coordinate, monitor, and evaluate the awareness rising campaigns and efforts against specific KPIs (e.g., effectiveness, usability, broadness, scope), and provide recommendations for corrective actions.

Intelligence building: Coordinate the effective collection, storage, and analysis of maritime cybersecurity information and international maritime intelligence.

Information sharing: Orchestrate and monitor the mutual sharing of all cybersecurity-related information (e.g., threats, vulnerabilities, market/civilian-triggered attacks/incidents, crisis warnings) among international maritime agencies (e.g., IMO, NATO, EMSA, CERTS, CSIRTS), analysis centers, law enforcement forces, and international intelligent services.

Management of maritime-specific and supply chain cyber risks: Coordinate the efforts among stakeholders (e.g., port authorities, maritime industry/manufacturers, insurance companies) to more effectively manage maritime cyber risks.

Next generation maritime research: Strengthen the prospects of maritime industrial markets by coordinating maritime military and civilian research efforts (e.g., programs, activities, funds) toward the facilitation of a common maritime cyber research agenda (which is continuously monitored and improved).

Maritime cybersecurity market: Coordinate and monitor actions (related to, e.g., certification, privacy labeling, accountability) to develop a trustworthy, maritime cybersecurity market.

BIBLIOGRAPHY

[1] A Digital Agenda for Europe (COM(2010)245), Brussels, August 26, 2010. Available at: http://eur-lex.europa.eu/LexUriServ/LexUriServ.do?uri=COM:2010:0245:FIN:EN:PDF.

[2] I. Abele-Wigert, M. Dunn, An inventory of 20 national and 6 international critical infrastructure protection policies, in: A. Wenger, V. Mauer (Eds.), International CIIP Handbook 2006, vol. I, Center for Security Studies, ETH Zurich, 2006.

[3] ACID (Atomicity, Consistency, Isolation, Durability): A Set of Properties of Database Transactions, 2017. https://en.wikipedia.org/wiki/ACID.

[4] R. Adler, J. Fuller, An integrated framework for assessing and mitigating risks to maritime critical infrastructure, in: Proc. of IEEE Conference on Technologies for Homeland Security, May 2007, pp. 252–257.

[5] C. Alberts, D. Audrey, Operationally Critical Threat, Asset, and Vulnerability Evaluation (OCTAVE) Method Implementation Guide, v2.0, Software Engineering Institute, Carnegie Mellon University, 2001. http://resources.sei.cmu.edu/library/asset-view.cfm?assetid=13473.

[6] AS/NZS 4360, Risk Management Standards Australia, Strathfield, 1999.

[7] Z.Z. Aung, K. Watanabe, A framework for modeling interdependencies in Japan's critical infrastructures, in: C. Palmer, S. Shenoi (Eds.), 3rd IFIP Int. Conf. on Critical Infrastructure Protection (CIP-2009), Springer, USA, 2009, pp. 243–257.

[8] J. Balmat, F. Lafont, R. Maifret, N. Pessel, MAritime RISk Assessment (MARISA), a fuzzy approach to define an individual ship risk factor, Ocean Engineering 36 (15–16) (November 2009) 1278–1286.

[9] BitNami Cloud Hosting for Open Source Applications, 2017. http://bitnami.org/cloud.

[10] J. Bohren, et al., Web Services Trust Language (WS-Trust), February 2005.

[11] BSI Standard 100-1, Information Security Management Systems (ISMS), 2005. Retrieved May 2008, from: www.bsi.bund.de.

[12] BSI Standard 100-2, IT-Grundszchutz Methodology, 2005. Retrieved from: www.bsi.bund.de.

[13] BSI Standard 100-3, Risk Analysis Based on IT-Grundszchutz, 2005. Retrieved from: www.bsi.bund.de.

[14] Bundesamt für Sicherheit in der Informationstechnik, IT-Grundschutz Kataloge, 2013. Online: https://www.bsi.bund.de/DE/Themen/ITGrundschutz/itgrundschutz_node.html.

[15] Cloud Computing Software from Eucalyptus, 2017. http://www.eucalyptus.com/.

[16] Club de la Securite de L'Information Francais Methods Commission, Mehari 2010 Risk Analysis and Treatment Guide, France, August 2010. Available at: http://www.clusif.asso.fr/fr/production/ouvrages/pdf/MEHARI-2010 Risk-Analysis-and-Treatment-Guide.pdf.

[17] Clusif Methods Commission, MEHARI V3 Risk Analysis Guide, 2004.

[18] Code of practice for Health and Safety in Port Operations, Occupational Safety and Health Service Department of Labour – PO Box 3705, Wellington, New Zealand, May 2004, ISBN: 0-478-28002-5.

[19] Collectd – The System Statistics Collection Daemon, 2017. https://collectd.org/.

[20] Common Attack Pattern Enumeration and Classification, 2017. https://capec.mitre.org/.

[21] Common Criteria Working Group, Common Methodology for Information Technology Security Evaluation – Evaluation Methodology, CCMB-2007-09-004 2007. http://www.commoncriteriaportal.org.

[22] Common Vulnerabilities and Exposures, 2017. https://cve.mitre.org/.

[23] Communication from the Commission of 12 December 2006 on a European Programme for Critical Infrastructure Protection (COM (2006) 786). Available at: http://eur-lex.europa.eu/LexUriServ/site/en/com/2006/com2006_0786en01.pdf.

[24] Communication from the Commission to the Council and the European Parliament on the Prevention, Preparedness and Response to Terrorist Attacks (COM(2004) 698), Brussels, October 20, 2004. http://eurlex.europa.eu/LexUriServ/LexUriServ.do?uri=COM:2004:0698:FIN:EN:PDF.

[25] Consistently Optimised Resilient Ecosystem, 2017. http://www.coreproject.eu/.

[26] Council Directive 2008/114/EC of 8 December 2008 on the Identification and Designation of European Critical Infrastructures and the Assessment of the Need to Improve Their Protection, 2017. http://eurlex.europa.eu/LexUriServ/LexUriServ.do?uri=OJ:L:2008:345:0075:0082:EN:PDF.

[27] F. Crespo, M. Gomez, J. Candau, J.A. Manas, MAGERIT –Version 2, Methodology for Information Systems Risk Analysis and Management, Books I – The Method, Ministerio de Administraciones Publicas, June 2006.

[28] CYSM European Commission, Programme Prevention, Preparedness and Consequence Management of Terrorism CIPS, 2012. http://www.cysm.eu/index.php/en/.

[29] R.-M. Darbra, J. Casal, Historical analysis of accidents in seaports, Department of Chemical Engineering, Centre d'Estudis del Risc Tecnològic (CERTEC), Universitat Politècnica de Catalunya, Safety Science 42 (2004) 85–98.

[30] S. De Porcellinis, G. Oliva, S. Panzieri, R. Setola, A holistic-reductionistic approach for modeling interdependencies, in: C. Palmer, S. Shenoi (Eds.), 3rd IFIP Int. Conf. on Critical Infrastructure Protection (CIP-2009), Springer, USA, 2009, pp. 215–227.

[31] K. Dellios, D. Papanikas, D. Polemi, Cyber ocean: a roadmap to maritime cloud, International Journal of Business Innovation and Research (IJBIR) (2014), Inderscience.

[32] B. Downs, The maritime security risk analysis model, in: US Coast Guard Proc. of the Marine Safety and Security Council, Spring 2007. Available at: http://www.uscg.mil/proceedings/.

[33] ENISA (European Network and Information Security Agency), Risk Management: Implementation Principles and Inventories for Risk Management/Risk Assessment Methods and Tools, 2017. https://www.enisa.europa.eu/publications/risk-management-principles-and-inventories-for-risk-management-risk-assessment-methods-and-tools.

[34] ENISA Report Cyber Security Aspects in the Maritime Sector, ENISA, 2011. http://www.enisa.europa.eu/activities/Resilience-and-CIIP/critical-infrastructure-and-services/dependencies-of-maritime-transport-to-icts.

[35] ENISA Workshop on Cyber Security Aspects in the Maritime Sector, Brussels, 2011. http://www.enisa.europa.eu/act/res/workshops-1/2011/cyber-security-aspects-in-the-maritime-sector.

[36] Environmental, Health, and Safety Guidelines for Ports, Harbors, and Terminals – International Finance Corporation – World Bank Group, April 2007.

[37] EU and NATO Cyber Defence Cooperation, 2017. http://collections.internetmemory.org/haeu/content/20160313172652/http://eeas.europa.eu/top_stories/2016/100216_eu-nato-cyber-defence-cooperation_en.htm.

[38] EU Benchmark Survey Confirms Member States Making Significant Progress in eGovernment, 2017. http://ec.europa.eu/information_society/newsroom/cf/itemlongdetail.cfm?item_id=3634.

[39] European Aviation Safety Agency. Available at: http://www.easa.europa.eu.
[40] European Commission, Regulation (EC) No 725/2004 of the European Parliament and of the Council of 31 March 2004 on enhancing ship and port facility security, Official Journal of the European Union (2004) 6–91, L 129/6.
[41] European Maritime Safety Agency. Available at: http://www.emsa.europa.eu/.
[42] Expression of Needs and Identification of Security Objectives PREMIER MINISTRE Secrétariat général de la défense nationale Direction centrale de la sécurité des systèmes d'information Sous-direction des opérations Bureau conseil, http://www.ssi.gouv.fr, 2017.
[43] eyeOS, http://www.eyeos.com/, 2017.
[44] FASTPASS, A Harmonized, Modular Reference System for All European Automated Border Crossing Points. https://www.fastpass-project.eu/.
[45] L.S. Fryer, R.F. Griffiths, Worldwide Data on the Incidence of Multiple-Fatalities Accidents, Safety and Reliability Directorate (SRD R 149), 1979.
[46] G. Giannopoulos, R. Filippini, M. Schimmer, Risk Assessment Methodologies for Critical Infrastructure Protection. Part I: A State of the Art, Publications Office of the European Union, Luxembourg, 2012. http://dx.doi.org/10.2788/22260. Joint Research Center Publication, JRC 70046, EUR 25286 EN, ISBN 978-92-79-23839-0, ISSN 1831-9424.
[47] P. Haastrup, L. Brockhoff, Severity of accidents with hazardous materials, Journal of Loss Prevention in Process Industries 3 (4) (1990) 395–405.
[48] R. Haavel, M. Oit, A. Usk, Maritime surveillance information availability in Estonia, in: E. Shahbazian et al. (Eds.), Harbor Protection through Data Fusion Technologies, Springer, Netherlands, 2009, pp. 53–60.
[49] Y. Haimes, J. Santos, K. Crowther, M. Henry, C. Lian, Z. Yan, Risk analysis in interdependent infrastructures, Critical Infrastructure Protection 253 (2007) 297–310.
[50] S. Holmberg, A systems perspective on supply chain measurements, International Journal of Physical Distribution and Logistics Management 30 (10) (2000) 847–868.
[51] ILO, Guidelines on occupational safety and health management systems, ILO-OSH 2001, in: Guide: Occupational Safety, Occupational Health, National Level, Enterprise Level, Technical Aspect, International Labour Office, Geneva, 2001, ISBN: 92-2-111634-4. 13.04.2.
[52] ILO, Safety and health in ports ILO code of practice, Geneva, in: Occupational Safety, Occupational Health, Port, Docker, Code of Practice, International Labour Office, 2005, ISBN: 92-2-115287-1. 13.04.2.
[53] INDG446 A Quick Guide to Health and Safety in Ports. Published by the Health and Safety Executive INDG446, ISBN: 978 0 7176 6421 4.
[54] Information Security Assessment & Monitoring Method (ISAMM), 2017. http://www.telindus.com.
[55] Insight Consulting, CRAMM User Guide, Issue 5.1, United Kingdom, 2005.
[56] International Convention for the Prevention of Pollution from Ships (MARPOL), 2017. http://www.imo.org/about/conventions/listofconventions/pages/international-convention-for-the-prevention-of-pollution-from-ships-%28marpol%29.aspx.
[57] International Convention for the Safety of Life at Sea (SOLAS), 1974, 2017. http://www.imo.org/about/conventions/listofconventions/pages/international-convention-for-the-safety-of-life-at-sea-%28solas%29,-1974.aspx.
[58] International Maritime Organisation, International Ship and Port Facility Security Code, London, United Kingdom, 2004.
[59] International Standardization Organization, ISO 20000: Information Technology Service Management, Geneva, Switzerland, 2005.
[60] International Standardization Organization, Ships and Marine Technology – Maritime Port Facility Security Assessments and Security Plan Development, Geneva, Switzerland, 2007.

[61] International Standardization Organization, ISO 28000: Specification for Security Management Systems for the Supply Chain, Geneva, Switzerland, 2007.
[62] International Standardization Organization, ISO 28001: Security Management Systems for the Supply Chain — Best Practices for Implementing Supply Chain Security, Assessments and Plans — Requirements and Guidance, Geneva, Switzerland, 2007.
[63] International Standardization Organization, ISO 31000: Risk Management – Principles and Guidelines, Geneva, Switzerland, 2009.
[64] International Standardization Organization, ISO 31010: Risk Management – Risk Assessment Techniques, Geneva, Switzerland, 2009.
[65] International Standardization Organization, ISO 27005: Information Security Risk Management, Geneva, Switzerland, 2011.
[66] International Standardization Organization, ISO 27001: Information Security Management System Requirements, Geneva, Switzerland, 2013.
[67] Inventory of Risk Management/Risk Assessment Tools, 2017. https://www.enisa.europa.eu/topics/threat-risk-management/risk-management/current-risk/risk-management-inventory/rm-ra-tools.
[68] ISO/IEC, Information Technology – Security Techniques – Code of Practice for Information Security Management, ISO/IEC 27002, 2005.
[69] ISO/IEC, Information Technology – Security Techniques – Specification for an Information Security Management System, ISO/IEC 27001, 2005.
[70] ISO/IEC, Information Technology – Security Techniques – Information Security Risk Management, ISO/IEC 27005, 2008.
[71] M. Jurczyk, One Font Vulnerability to Rule Them All #1: Introducing the BLEND Vulnerability, 2017. Google Project Zero. https://googleprojectzero.blogspot.gr/2015/07/one-font-vulnerability-to-rule-them-all.html.
[72] B. Kakish, Harbour protection in the Jordanian port of Aqaba, in: E. Shahbazian et al. (Eds.), Harbour Protection Through Data Fusion Technologies, Springer, Netherlands, 2009, pp. 37–41.
[73] E.-M. Kalogeraki, N. Polemi, S. Papastergiou, A modeling approach for maritime supply chain services, in: 12th Edition of the International Forum on Knowledge Asset Dynamics: Knowledge Management in the 21st Century: Resilience, Creativity and Co-creation (IFKAD2017), 2017.
[74] E.-M. Kalogeraki, N. Polemi, S. Papastergiou, T. Panayiotopoulos, Modelling SCADA attacks, in: World Conference on Smart Trends in Systems, Security and Sustainability (WS4 2017), Conference Proceedings, ISSN: 2367-3370. Springer Lecture Notes in Networks and Systems (LNNS), February 2017, pp. 15–16 (London, England).
[75] A. Karantjias, N. Polemi, An innovative platform architecture for complex secure e/m government services, Special Issue on Mobile Services Technological and Legal Issues, Inderscience Publishers, International Journal of Electronic Security and Digital Forensics (IJESDF) 2 (4) (2009) 338–354.
[76] A. Karantjias, N. Polemi, S. Papastergiou, Advanced security management system for critical infrastructures, in: Special Session on "Secure and Sustainable Maritime Digital Environment" Within the Fifth International Conference on Information, Intelligence, Systems and Applications (IISA 2014), July 07, 2014 (Chania Crete, Greece).
[77] P. Kotzanikolaou, M. Theoharidou, D. Gritzalis, Assessing n-order dependencies between critical infrastructures, International Journal Critical Infrastructures Protection 9 (1/2) (2013).
[78] Liberty Alliance Project White Paper: Liberty Alliance & WS-Federation: A Comparative Overview, October 2003. http://www.projectliberty.org/liberty/content/download/402/2765/file/wsfed-liberty-overview-10-13-03.pdf.

[79] H. Lockhart, et al., Web Services Federation Language (WS-Federation), Version 1.1, December 2006.
[80] E. Luiijf, H. Burger, M. Klaver, Critical infrastructure protection in the Netherlands: a quick-scan, in: U. Gattiker, P. Pedersen, K. Petersen (Eds.), In Proc. of EICAR Conference, 2003.
[81] G. Makridimitris, D. Polemi, C. Douligeris, Security risk assessment challenges in port information technology systems, in: Volume 441 of the Communications in Computer and Information Science Series, 2014.
[82] Managing Health and Safety. Five Steps to Success, Published by the Health and Safety Executive INDG275, ISBN:978 0 7176 2170 5.
[83] MariaDB, Database Servers, 2017. https://mariadb.org/.
[84] Maritime Domain Awareness Data Sharing Community of Interest (MDA DS COI), Data Management Working Group, Spiral 2, Vocabulary Handbook Version 2.0.2, November 15, 2007. Available at: http://www.uscg.mil/acquisition/nais/RFP/SectionJ/MDA-COI-vocab.pdf.
[85] MEDUSA Multi-Order Dependency Approaches for Managing Cascading Effects in Ports' Global Supply Chain and Their Integration in Risk Assesment Frameworks. http://medusa.cs.unipi.gr/.
[86] MEDUSA, Multi-order dependency approaches for managing cascading effects in ports' global supply chain and their integration in risk assesment frameworks, in: European Commission, Programme Prevention, Preparedness and Consequence Management of Terrorism CIPS, 2014. http://athina.cs.unipi.gr/medusa/.
[87] MHIDAS, Mayor Hazard Incident Data Service (OHS_ROM, Reference Manual), 2002.
[88] Ministry of the Interior and Kingdom Relations, National Risk Assessment Method Guide 2008, National Security Programme, Netherlands, June 2008. Available at: http://www.minbzk.nl//bzk2006uk/subjects/public-safety/publications/115647/national risk/.
[89] [MITIGATE D2.2] MITIGATE Deliverable 2.2, Evidence-Driven Maritime Supply Chain Risk Assessment Approach.
[90] [MITIGATE D3.1] MITIGATE Deliverable 3.1, Specification of the Risk Assessment, Mitigation and Simulation Functionalities, 2016.
[91] [MITIGATE D3.2] MITIGATE Deliverable 3.2, Prediction, Forecasting, Visualization and Open Intelligence Services Specification, 2016.
[92] MITIGATE: (Multidimensional, Integrated, Risk Assessment Framework and Dynamic, Collaborative Risk Management Tools for Critical Information Infrastructures), 2017. http://www.mitigateproject.eu/.
[93] MongoDB, 2017. https://www.mongodb.com/.
[94] National Institute for Standards and Technology, Risk Management Guide for Information Technology Systems, NIST Special Publication 800–830, USA, July 2002.
[95] Neo4j, https://neo4j.com/, 2017.
[96] New Zealand Port and Harbour Marine Safety Code, Maritime Safety Authority of New Zealand – Kia Maanu Kia Ora, Copyright Maritime Safety Authority, 2004.
[97] Nimbus, 2017. http://www.nimbusproject.org/.
[98] [NIST NVD] National Institute of Standards and Technology, National Vulnerability Database, 2012. Retrieved from: http://nvd.nist.gov/cvss.cfm.
[99] [NIST Security Considerations in SDLC] S. Radack, Security Considerations in the Information System Development Life Cycle, National Institute of Standards and Technology, 2008. Retrieved from: http://www.itl.nist.gov/lab/bulletns/bltndec03.htm.
[100] [NIST SP800-16] M. Wilson, K. Stine, P. Bowen, Information Security Training Requirements: A Role- and Performance-Based Model, NIST Special Publication 800-16, revision 1.0. National Institute of Standards and Technology, 2009. Retrieved from: http://csrc.nist.gov/publications/drafts/800-16-rev1/Draft-SP800-16-Rev1.pdf.

[101] [NIST SP800-37] National Institute of Standards and Technology, Joint Task Force Transformation Initiative, Guide for Applying the Risk Management Framework to Federal Information Systems, NIST Special Publication 800-37, 2010. Retrieved from: http://csrc.nist.gov/publications/nistpubs/800-37-rev1/sp800-37-rev1-final.pdf.

[102] [NIST SP800-40] P. Mell, T. Bergeron, D. Henning, Creating a Patch Management and Vulnerability Management Program, NIST Special Publication 800-40, version 2.0 National Institute of Standards and Technology, 2005. Retrieved from: http://csrc.nist.gov/publications/nistpubs/800-40-Ver2/SP800-40v2.pdf.

[103] [NIST SP800-53] National Institute of Standards and Technology, Joint Task Force Transformation Initiative, Recommended Security Controls for Federal Information Systems and Organizations, NIST Special Publication 800-53, revision 3, 2009. Retrieved from: http://csrc.nist.gov/publications/nistpubs/800-53-Rev3/sp800-53-rev3-final_updated-errata_05-01-2010.pdf.

[104] [NIST SP800-61] K. Scarfone, T. Grance, K. Masone, Computer Security Incident Handling Guide, NIST Special Publication 800-61, revision 1, National Institute of Standards and Technology, 2008. Retrieved from: http://csrc.nist.gov/publications/nistpubs/800-61-rev1/SP800-61rev1.pdf.

[105] [NIST SP800-64] R. Kissel, K. Stine, M. Scholl, H. Rossman, J. Fahlsing, J. Gulick, Security Considerations in the System Development Life Cycle, NIST Special Publication 800-64, revision 2, National Institute of Standards and Technology, 2008. Retrieved from: http://csrc.nist.gov/publications/nistpubs/800-64-Rev2/SP800-64-Revision2.pdf.

[106] [NIST SP800–128] National Institute of Standards and Technology, Guide for Security-Focused Configuration Management of Information Systems, Special Publication 800-128, 2011. Retrieved from: http://csrc.nist.gov/publications/nistpubs/800-128/sp800-128.pdf.

[107] [NIST SP800-137] K. Dempsey, N.S. Chawla, A. Johnson, R. Johnston, A.C. Jones, A. Orebaugh, K. Stine, Information Security Continuous Monitoring (ISCM) for Federal Information Systems and Organizations, NIST Special Publication 800-137. National Institute of Standards and Technology, 2011. Retrieved from: http://csrc.nist.gov/publications/nistpubs/800-137/SP800-137-Final.pdf.

[108] [NISTIR 7622] M. Swanson, N. Bartol, R. Moorthy, Piloting Supply Chain Risk Management for Federal Information Systems, Draft NISTIR 7622. National Institute of Standards and Technology, 2010. Retrieved from: http://csrc.nist.gov/publications/drafts/nistir-7622/draft-nistir-7622.pdf.

[109] [NISTIR 7628] The Smart Grid Interoperability Panel – Cyber Security Working Group, Guidelines for Smart Grid Cyber Security: Vol. 1, Smart Grid Cyber Security Strategy, Architecture, and High-Level Requirements (NISITIR 7628), National Institute of Standards and Technology, 2010. Retrieved from: http://csrc.nist.gov/publications/nistir/ir7628/nistir-7628_vol1.pdf.

[110] North American Reliability Corporation (NERC). Available at: http://www.nerc.com.

[111] S. Novikov, Implementation of the ISPS code in the Russian federation: ships and ports, in: E. Shahbazian et al. (Eds.), Harbour Protection Through Data Fusion Technologies, Springer, Netherlands, 2009, pp. 23–26.

[112] T. Ntouskas, N. Polemi, A secure, collaborative environment for the security management of port information systems, in: Proceedings of the Fifth International Conference on the Internet and Web Applications and Services, ICIW 2010, IEEE Computer Society Digital Library, Barcelona, Spain, 2010, pp. 374–379.

[113] T. Ntouskas, N. Polemi, Collaborative Security Management Services for Port Information Systems, DCNET/ICE-B/OPTICS, 2012, pp. 305–308.

[114] Oasis Standard, Security Assertion Markup Language (SAML) V2.0, 2005.

[115] OCTAVE Method Implementation Guide Version 2.0, Carnegie Mellon University, June 2001. Available at: http://www.cert.org/octave/, 2010.
[116] [OECD Reducing Systemic Cybersecurity Risk] P. Sommer, I. Brown, Reducing Systemic Cybersecurity Risk, Organisation for Economic Co-operation and Development, 2011. Retrieved from: http://www.oecd.org/dataoecd/57/44/46889922.pdf.
[117] Open Source Software for Building Private and Public Clouds, 2017. http://openstack.org/.
[118] OpenNebula: The Open Source Solution for Data Center Virtualization, 2017. http://opennebula.org/.
[119] S. Papastergiou, N. Polemi, Harmonizing commercial port security practices & procedures in mediterranean basin, in: SSMDE: Secure and Sustainable Maritime Digital Environment, IISA 2014, Springer Verlag, 2014.
[120] S. Papastergiou, N. Polemi, MITIGATE: a dynamic supply chain cyber risk assessment methodology, in: World Conference on Smart Trends in Systems, Security and Sustainability (WS4 2017), Springer Computer Science Proceedings, February 2017, pp. 15–16 (London, England).
[121] S. Papastergiou, N. Polemi, I. Papagiannopoulos, Business and threat analysis of ports' supply chain services, in: Special Session on "Innovative Risk Management Methodologies and Tools for Critical Information Infrastructures (CII)" within the 6th International Conference on Digital Human Modeling and Applications in Health, Safety, Ergonomics and Risk Management (HCI International 2015), 2–7 August 2015, Los Angeles, CA, USA, 2015.
[122] S. Papastergiou, N. Polemi, A. Karantjias, CYSM: an innovative physical/cyber security management system for ports, in: Special Session on "Innovative Risk Management Methodologies and Tools for Critical Information Infrastructures (CII)" within the 6th International Conference on Digital Human Modeling and Applications in Health, Safety, Ergonomics and Risk Management (HCI International 2015), 2–7 August, 2015, Los Angeles, CA, USA, 2015.
[123] P. Pederson, D. Dudenhoeffer, S. Hartley, M. Permann, Critical Infrastructure Interdependency Modeling: A Survey of U.S. And International Research, INL, INL/EXT-06-11464, 2006.
[124] T.R. Peltier, Information Security Risk Analysis, Auerbach Publications, 2001.
[125] E. Planas-Cuchi, H. Montiel, J. Casal, A survey of the origin, type and consequences of fire accidents in process plants and in the transportation of hazardous material, Transaction IchemE 75 (1997) 3–8.
[126] E. Planas-Cuchi, J.A. Vílchez, F.X. Pérez-Alavedra, J. Casal, Effects of fire on a container storage system— a case study, Journal of Loss Prevention in Process Industries 11 (5) (1998) 323–331.
[127] N. Polemi, Security Management of the Ports' Information Systems. ENISA Personal Study. Available at: http://www.enisa.europa.eu (accessed August 2012).
[128] N. Polemi, P. Kotzanikolaou, Medusa: a supply chain risk assessment methodology, in: Cyber Security and Privacy Forum (CSP2015), Springer International Publishing, 2015, pp. 79–90.
[129] N. Polemi, P. Kotzanikolaou, Medusa: a supply chain risk assessment methodology, CSP forum, in: Cyber Security and Privacy Innovation Forum, 28–29/4/15, Springer Verlag, 2015. Lecture Notes. https://www.cspforum.eu/2015.
[130] N. Polemi, P. Kotzanikolaou, S. Papastergiou, Design and Validation of the MEDUSA Supply Chain Risk Assessment Methodology and System, Elsevier International Journal of Critical Infrastructure Protection (IJIP), 2016.
[131] N. Polemi, T. Ntouskas, Open issues and proposals in the IT security management of commercial ports: the S-PORT national case, in: D. Gritzalis, S. Furnell, M. Theoharidou (Eds.), 2012, pp. 567–572. SEC 2012, IFIP AICT 376.

[132] D. Polemi, T. Ntouskas, E. Georgakakis, C. Douligeris, M. Theoharidou, D. Gritzalis, S-port: collaborative security management of port information systems, in: Proc. of the 4th International Conference on Information, Intelligence, Systems and Applications (IISA-2013), IEEE Press, Greece, July 2013.
[133] D. Polemi, S. Papastergiou, Current efforts in ports and supply chains risk assessment, in: IEEE Proceedings of the 10th International Conference for Internet Technologies and Secure Transactions, London, U.K, 2015.
[134] Proposal for a Directive of the European Parliament and of the Council Concerning Measures to Ensure a High Level of Network and Information Security Across the Union (SWD(2013)), February 7, 2013. Available at: http://eeas.europa.eu/policies/eu-cyber-security/cybsec_impact_ass_res_en.pdf.
[135] Protecting Europe from Large Scale Cyber-Attacks and Disruptions: Enhancing Preparedness, Security and Resilience, COM (2009)149 of 30 March 2009. Available at:http://eur-lex.europa.eu/LexUriServ/LexUriServ.do?uri=COM:2009:0149:FIN:EN:PDF.
[136] D. Reynolds, K. Rabey, N. Polemi, Analysing Mes Needs and Expectations in the Area of Information Security, ENISA report 2008. Available at: http://www.enisa.europa.eu/act/sr/reports/micro-enterprises/files/wg-micro-report.
[137] S.M. Rinaldi, J.P. Peerenboom, T.K. Kelly, Identifying, understanding and analyzing critical infrastructure interdependencies, IEEE Control Systems (December 2001) 11–25.
[138] H. Romer, L. Brockhoff, P. Haastrup, H.J. Styhr Petersen, Marine transport of dangerous goods. Risk assessment based on historical accident data, Journal of Loss Prevention in Process Industries 6 (4) (1993) 219–225.
[139] http://www.rsa.com/rsalabs/node.asp?id=2118.
[140] SAFESEANET, A European Platform for Maritime Data Exchange Between Member States' Maritime Authorities, is a Network/Internet Solution Based on the Concept of a Distributed Database. http://ec.europa.eu/idabc/en/document/2282/5926.html.
[141] SafeSeaNet. http://www.emsa.eu.int/.
[142] [SCADA AU RMF] IT Security Expert Advisory Group, Generic SCADA Risk Management Framework for Australian Critical Infrastructure, 2012. Retrieved from: http://www.tisn.gov.au/Documents/SCADA-Generic-Risk-Management-Framework.pdf.
[143] [SEI CMM] M. Paulk, C. Weber, S. Garcia, M.B. Chrissis, M. Bush, Key Practices of the Capability Maturity Model, Version 1.1, Technical Report CMU/SEI-93-TR-25, Software Engineering Institute, Carnegie Mellon University, 1993. Retrieved from: http://www.sei.cmu.edu/reports/93tr025.pdf.
[144] [Situation Awareness in Dynamic Systems] M. Endsley, Toward a theory of situation awareness in dynamic systems, Human Factors (1995) 32–64.
[145] A. Sokolov, Steps to better waterside port and harbor security: the development of regional maritime safety systems in the Russian federation, in: E. Shahbazian et al. (Eds.), Harbour Protection Through Data Fusion Technologies, Springer, Netherlands, 2009, pp. 27–32.
[146] S-PORT, A Secure, Automated Collaborative Environment for the Creation of Risk Assessment Methodologies, Generation of Business Continuity and Disaster Recovery Plans for the Port Information Systems. Funded by the GSRT (General Secretariat for Research and Technology of the Ministry of Development), Greek National Research Programme "COOPERATION" (NSRF 2007–2013). http://s-port.unipi.gr/index.php/.
[147] G. Stoneburner, A. Goguen, A. Feringa, Risk Management Guide for Information Technology Systems, Special Publication 800–830, National Institute of Standards and Technology, 2002.

[148] [Supply Chain Risk Management Awareness] J. Filsinger, B. Fast, D.G. Wolf, J.F.X. Payne, M. Anderson, Supply Chain Risk Management Awareness, Armed Forces Communication and Electronics Association Cyber Committee, 2012. Retrieved from: http://www.afcea.org/committees/cyber/documents/Supplychain.pdf.
[149] The International Maritime Organization, http://www.imo.org, 2017.
[150] The Maritime Security Council (MSC), http://www.maritimesecurity.org/, 2017.
[151] The Stationery Office (TSO), Continual Service Improvement, ITIL V3, 2007.
[152] M. Theoharidou, P. Kotzanikolaou, D. Gritzalis, Risk assessment methodology for interdependent critical infrastructures, International Journal of Risk Assessment and Management 15 (2/3) (2011).
[153] Thymeleaf: A Modern Server-Side Java Template Engine for Both Web and Standalone Environments, 2017. http://www.thymeleaf.org/.
[154] Trans-European Transport Network Executive Agency. http://tentea.ec.europa.eu/.
[155] U.S. Department of Energy, Resource Handbook on DOE Transportation Risk Assessment, Report DOE/EM/NTP/HB-01, National Transportation Program, Office of Environmental Management, USA, 2002.
[156] Ubuntu. Available at: http://www.ubuntu.com/cloud.
[157] UNCTAD, Review of Maritime Transport 2012, New York and Geneva, 2012.
[158] J.A. Vílchez, S. Sevilla, H. Montiel, J. Casal, Historical analysis of accidents in chemical plants and in the transportation of hazardous materials, Journal of Loss Prevention in Process Industries 2 (8) (1995) 87–96.
[159] J.M. Yusta, G.J. Correa, R. Lacal-Arántegui, Methodologies and applications for critical infrastructure protection: state-of-the-art, ELSEVIER, Energy Policy 39 (10) (October 2011) 6100–6119.
[160] E. Zambon, S. Etalle, R.J. Wieringa, P.H. Hartel, Model-based qualitative risk assessment for availability of IT infrastructures, Software and Systems Modeling 10 (4) (2011) 553–580.
[161] E. Zio, G. Sansavini, Modeling interdependent network systems for identifying cascade-safe operating margins interdependency, IEEE Transactions on Reliability 60 (1) (March 2011).

APPENDIX A

CYSM QUESTIONNAIRE FOR PORTS' SECURITY AWARENESS

Questionnaire for Ports and Port Operators on Security Awareness

Information (Optional)

Company	______________________________
Name	______________________________
Position	______________________________
Number of employees	______________________________
Business activity	______________________________
URL	______________________________
e-mail	______________________________

Dear Port Operators,

In the context of Deliverable 2.1 of the CYSM Project, a survey will be run to identify the security systems that the port operators are using.

The objective of this questionnaire is to understand the security awareness level of the ports and their critical ICT infrastructure in terms of threats and vulnerability of all layers of the ports' ICT systems, i.e., (a) physical infrastructure, (b) telco infrastructure equipment, (c) software, (d) services and applications, (e) data, and (f) users (external/internal persons and things). In addition, an overview of the services and users vis-à-vis vulnerability will be sought.

The questionnaire intends to identify whether:

- port operators have put in place a security plan (port security plan and ICT security policy), in which extent it is applied for, and which objectives it has;
- port employees and users are aware of the security issues related with their daily activities;
- information security awareness training is offered as well as its effectiveness;
- interrelationships between competent authorities and ports exist and for which reasons.

For the purposes of this survey, only the ports under their capacity as operators are considered (i.e., ports that are simultaneously landlords and operators OR terminal operators).

The responses will be considered confidential and only aggregate figures will be reported.

Port Security Plan

1. Does the port offer a well-defined security management plan?

☐ Yes: go to questions 1a

☐ No: go to question 2

1a. Which model for security management do you adopt? [Choose only one.]

☐	In house, e.g., from a specific dept. or person(s) from the port (e.g., PFSO)
☐	Outsource, e.g., other company or organization or external expert(s)
☐	Other

1b. Which better describes your perception of security? [Choose only one.]

☐	Does not concern/influence your business activities
☐	It is not your first priority but you plan to consider it in the future
☐	You understand its necessity, but it is too expensive to consider it

1c. Which security schemes does your security plan cover? [Choose as many as appropriate.]

☐	ISPS
☐	ISO 9001
☐	ISO 17779
☐	ISO 27001
☐	ISO 28000
☐	Other (which: ________________________________)
☐	Other (which: ________________________________)
☐	Other (which: ________________________________)

1d. Does your security scheme identify and analyze the following in depth?

Activities	Our security plan does not cover this	Our security plan covers it, and... Stakeholders present valid documents, but we do not analyze them	Our security plan covers it, and... We keep an active database of them and have specific detailed processes for analyses
Ship security plans	☐	☐	☐
Ship security officers	☐	☐	☐
Company security officers	☐	☐	☐
Onboard equipment	☐	☐	☐
Corresponding port facility security plans	☐	☐	☐
Port facility security officers	☐	☐	☐
Certain security equipment	☐	☐	☐
Certain security features	☐	☐	☐
Monitoring the activities of people and cargo	☐	☐	☐
Secure communications	☐	☐	☐
Secure access in port sectors (monitoring, control)	☐	☐	☐
Cyber port policy	☐	☐	☐
Other (please specify)	☐	☐	☐
Do not know		☐	

2. Did you consider security issues in the following activities of the port in the past 12 months? If yes, were your reactions pro-active or reactive?

Activities	Security issues were not considered	Yes, security issues were considered	
		Nature of action	
		Proactive	Reactive
Improvements in your services/ products/business	☐	☐	☐
Industry compliance	☐	☐	☐
Improvements in the operation of your IT systems	☐	☐	☐
Improvements in the operation of your port facilities entry/exit/ perimeter systems	☐	☐	☐
Improvements in your business profile and confidence level of your users and services	☐	☐	☐
Compliance with national legislation influencing your operation and business	☐	☐	☐
Privacy and data protection	☐	☐	☐
Safety of personnel and users	☐	☐	☐
Secure access in port sectors	☐	☐	☐
Do not know	☐		

3. Select three (3) main categories of regulations/legislation that influence your operation and security measures in the past/ next 12 months. Indicate if you consider any security projects in your selections.

Types of regulations	In the last 12 months	In the next 12 months	Security projects in progress/under consideration
Internal audit:			
ISPS	☐	☐	☐
ISM	☐	☐	☐
ISO families of standards	☐	☐	☐

Types of regulations	In the last 12 months	In the next 12 months	Security projects in progress/under consideration
OHSAS 18000 families	☐	☐	☐
Sarbanes-Oxley	☐	☐	☐
Regulation (EC) 725/2004	☐	☐	☐
SOLAS/MARPLE standards	☐	☐	☐
Council Directive 2008/114/EC on the identification and designation of European critical infrastructures and the assessment of the need to improve their protection.			
Data protection (e.g., Data Protection Act, Privacy in Telecoms, ISO 27000/27001/27002)	☐	☐	☐
Sectored regulations (e.g., PCI Data Security Standard)	☐	☐	☐
Customs and financial authorities' regulations	☐	☐	☐
Other (please define_________________________)	☐	☐	☐
Other (please define_________________________)	☐	☐	☐
Other (please define_________________________)	☐	☐	☐

4. **Select five (5) security issues in terms of importance (1 = low importance, 5 = highest importance) for your enterprise. Do you have a plan for addressing these security issues?**

Security issues	Important security issue	Plan of response		
		Now	Next 12 months	No plan
Secure data dissemination (to interested parties, to competent authorities, etc.)	☐	☐	☐	☐
Secure storage of data	☐	☐	☐	☐
Access (physical) to/from port facilities	☐	☐	☐	☐

Security issues	Important security issue	Plan of response		
		Now	Next 12 months	No plan
Secure port services	☐	☐	☐	☐
Radio frequency identifiers (RFID)	☐	☐	☐	☐
Mobile devices (e.g., PDA, smart phones)	☐	☐	☐	☐
Automated border control (ABCs) systems	☐	☐	☐	☐
Mobile storage (e.g., USB flash drive, portable drives)	☐	☐	☐	☐
Voice-over-IP telephony	☐	☐	☐	☐
Web services	☐	☐	☐	☐
Wireless networks	☐	☐	☐	☐
Secure satellite communication	☐	☐	☐	☐
New operation systems	☐	☐	☐	☐
Cryptography of e-mails	☐	☐	☐	☐
Cryptography of hard drive	☐	☐	☐	☐
	☐	☐	☐	☐

Security Practices

5. How do you check the security level of your enterprise? [Choose more than one if necessary.]

☐ Internal audit (from security department)

☐ External audit

☐ Internal assessment from personnel

☐ Other

☐ We do not have such procedures

6. Select five (5) main security activities in terms of importance (1 = low importance, 5 = highest importance) for your port. Independently select five (5) main security activities in terms of required implementation time (1 = short time consuming, 5 = high time consuming):

	Importance	Implementation time
Compliance with ISP	☐	☐
Compliance with ISO27001, 27005	☐	☐
Business continuity management	☐	☐
Risk and security management	☐	☐
Pursue, development, integration of IT systems	☐	☐
Access–control management	☐	☐
Communications and operations management	☐	☐
Physical security	☐	☐
Protection of employees (safety)	☐	☐
Updates of services and products	☐	☐
Security management	☐	☐
(Cyber) security policy	☐	☐
Privacy data protection	☐	☐
Accountability	☐	☐
Disaster recovery plan	☐	☐

7. How do you approach/solve the following security issues in your port?

Security issues	No consideration	Best practices	Own methods
Cyber security incidents, attacks	☐	☐	☐
Business continuity	☐	☐	☐
Business resilience	☐	☐	☐
Cyber risk management	☐	☐	☐
Physical risk management	☐	☐	☐
Privacy and data protection	☐	☐	☐
Security mechanisms in offered e/m port services	☐	☐	☐

8. When you collaborate with other enterprises (e.g., outsourcing, external contractors, third parties), which of the following do you consider necessary with respect to security?

- ☐ Your collaborators have and follow an auditable security policy
- ☐ Your collaborators are required to use/respect your own security policy and security concerns
- ☐ Your collaborators are ISO 27001 compliant
- ☐ None of the above
- ☐ Other

9. Have you ever performed a physical risk assessment?

- ☐ Yes
- ☐ No

10. How often do you report on the physical security risks and incidents and to whom?

	Frequency				
	Daily	Monthly	Trimester	Semester	Never
Port facility security officer	☐	☐	☐	☐	☐
Security company (external)	☐	☐	☐	☐	☐
Police	☐	☐	☐	☐	☐
Customs	☐	☐	☐	☐	☐
Competent authorities (which: ________________)	☐	☐	☐	☐	☐
Insurance companies	☐	☐	☐	☐	☐
Banks and financial institutions	☐	☐	☐	☐	☐
Contractors and stakeholders	☐	☐	☐	☐	☐
Chamber of commerce (or similar organization)	☐	☐	☐	☐	☐
National certification authority	☐	☐	☐	☐	☐
Other (which ____________)	☐	☐	☐	☐	☐
Other (which ____________)	☐	☐	☐	☐	☐
Other (which ____________)	☐	☐	☐	☐	☐

11. Have you ever performed a cyber risk assessment?

- ☐ Yes
- ☐ No

12. How often do you report on the cyber security risks and incidents and to whom?

	Frequency				
	Daily	Monthly	Trimester	Semester	Never
Port cyber security officer	☐	☐	☐	☐	☐
Security company (external)	☐	☐	☐	☐	☐
Public authorities (e.g., police)	☐	☐	☐	☐	☐
Maritime company	☐	☐	☐	☐	☐
Competent authorities (which: ____________)	☐	☐	☐	☐	☐
Insurance companies	☐	☐	☐	☐	☐
Banks and financial institutions	☐	☐	☐	☐	☐
Contractors and stakeholders	☐	☐	☐	☐	☐
IMO, EMSA (or similar maritime organization)	☐	☐	☐	☐	☐
National certification authority	☐	☐	☐	☐	☐
Other (which ____________)	☐	☐	☐	☐	☐
Other (which ____________)	☐	☐	☐	☐	☐
Other (which ____________)	☐	☐	☐	☐	☐

13. How do you get informed about security issues? Which organization would be most appropriate to get training on security? Which topics would you be most interested? (You may make several choices.)

Organizations	Security technologies/ implementation	Security policies	Risk assessment	Privacy, data protection
Security consultants (private companies)	☐	☐	☐	☐
University	☐	☐	☐	☐
National body	☐	☐	☐	☐
Internet	☐	☐	☐	☐
Governmental body	☐	☐	☐	☐
Chambers of commerce	☐	☐	☐	☐
Industry associations or professional bodies (e.g., IMO, EMSA)	☐	☐	☐	☐
Press, scientific publications	☐	☐	☐	☐

14. Which security standards are you aware of?

ISO 9001	☐
ISPS	☐
ISM	☐
SIRE/ATB Inspections	☐
ISA/IEC-62443	☐
US Barge and Inland ATB Inspection Request	☐
EBIS Inspection	☐
ISO/IEC 17799:2005	☐
ISO 20858	☐
ISO/IEC 27001	☐
Information Security Forum	☐

CobIT	☐
Information Technology Infrastructure Library (ITIL)	☐
Capability Maturity Model Integration (CMMI)	☐
SEISMED	☐
None of the above	☐
Other (which______________________________)	☐
Other (which______________________________)	☐
Add more if necessary	

15. What is the status of your port with respect to the following security activities?

Activity	In place/in progress	In place in the next 12 months	Not planned
Physical access using smart cards	☐	☐	☐
Physical access using paper-based ID cards	☐	☐	☐
Physical access using automated border control (ABC) systems	☐	☐	☐
Use of different authentication technologies for physical access (e.g., proximity cards, smart cards, biometrics)	☐	☐	☐
Use of intrusion detections systems	☐	☐	☐
Up-to-date antivirus and antispam filters	☐	☐	☐
Use of firewalls	☐	☐	☐
Inspections (out of schedule) on ships	☐	☐	☐

Thank you for your time

APPENDIX B

THREAT ANALYSIS: AN EXAMPLE

Asset type: Physical infrastructure

Threats	Vulnerabilities
Fire	Availability of flammable materials such as paper or boxes Backup files and systems not available Lack of fire detection devices Lack of physical security Location is in an area susceptible to natural disasters No business continuity plans or procedures for recovery of information and information assets
Earthquake	Backup files and systems not available Location is in an area susceptible to natural disasters No business continuity plans or procedures for recovery of information and information assets
Flood	Backup files and systems not available Location is in an area susceptible to natural disasters No business continuity plans or procedures for recovery of information and information assets
Storm	Backup files and systems not available Location is in an area susceptible to natural disasters No business continuity plans or procedures for recovery of information and information assets
Tidal surge/wave	Backup files and systems not available Location is in an area susceptible to natural disasters No business continuity plans or procedures for recovery of information and information assets
Contamination	Backup files and systems not available Lack of maintenance of equipment and facilities Location is in an area susceptible to environmental conditions such as contamination, electronic interference extreme temperature, and humidity vermin No business continuity plans or procedures for recovery of information and information assets
Electronic interference	Backup files and systems not available No business continuity plans or procedures for recovery of information and information assets

Continued

—Continued

Threats	Vulnerabilities
Extremes of temperature and humidity	Backup files and systems not available Location is in an area susceptible to environmental conditions such as contamination, electronic interference extreme temperature, and humidity vermin No business continuity plans or procedures for recovery of information and information assets Inadequate physical and environmental security policy and procedures Lack of a uniform physical security policy enforcement Lack of environmental protection Inadequate monitoring of environmental conditions Inadequate recovery procedure No formal or informal disaster or recovery plans No concrete assignment of continuity or disaster-related roles and responsibilities Inadequate change management procedure for infrastructure components Improper or inappropriate maintenance of technical facilities Location is in an area susceptible to environmental conditions such as extreme temperature and humidity Lack of backup facilities or processes Inadequate data backup procedure for both software and data Inadequate backup policy
Power fluctuations	Backup files and systems not available No business continuity plans or procedures for recovery of information and information assets Inadequate physical and environmental security policy and procedures Lack of a uniform physical security policy enforcement Lack of environmental protection Inadequate monitoring of environmental conditions Location is in an area susceptible to power fluctuations No uninterruptible power supply equipment No power conditioning equipment Inadequate change management procedure for infrastructure components Improper or inappropriate maintenance of technical facilities

—*Continued*

Threats	Vulnerabilities
Vermin (adware, malware, phishing, pop-ups, spyware, viruses, trojans, and worms)	Backup files and systems not available No business continuity plans or procedures for recovery of information and information assets
Failure of outsourced operations	Backup files and systems not available No business continuity plans or procedures for recovery of information and information assets
Transmission errors	Backup files and systems not available No business continuity plans or procedures for recovery of information and information assets Lack careful planning and laying of cables Lack of properly operation of network equipment Lack of cryptographic means to protect integrity of data
Unauthorized software changes	Backup files and systems not available
Technical failures	Lack of backup facilities or processes
Denial of service	Inadequate network management (resilience of routing)
Unauthorized dial-in access	Lack of a firewall
Unauthorized data access	Lack of a firewall Lack of physical security
Web site intrusion	Incorrectly configured or maintained security safeguards
Theft and fraud	Inadequate physical and environmental security policy and procedures No documented policies and procedures for physical control of hardware and software Inadequate physical security Lack of logical access security Lack of a uniform policy and procedure for controlling physical access to work areas and hardware (computers, communication devices, etc.) and software media enforcement Inadequate monitoring of the organization premises Inadequate audit logs to detect unauthorized access of the premises Lack of a formal entitlement review process regarding the access rights of the employees in the organization's premises Inadequate maintenance of the records regarding the repairs and modifications of the organization facilities physical components Inadequate change management procedure for infrastructure components

Continued

—*Continued*

Threats	Vulnerabilities
	No documented and tested security plans for safeguarding the systems and networks Lack of a comprehensive security awareness and training program Insufficient security training No concrete assignment of security roles and responsibilities Lack of physical security
Industrial action	Lack of an industrial agreement
Malicious destruction of data and facilities	Lack of physical security
Sabotage	Lack of physical security
Terrorist attacks	Inadequate physical and environmental security policy and procedures Lack of a uniform physical security policy enforcement Lack of environmental protection No business continuity plans for recovery of information and information assets Inadequate recovery procedure No formal or informal disaster or recovery plans No concrete assignment of continuity or disaster-related roles and responsibilities Improper or inappropriate maintenance of technical facilities Lack of backup facilities or processes Inadequate data backup procedure for both software and data Inadequate backup policy Backup files and systems not available Inadequate physical security Lack of logical access security Lack of a uniform policy and procedure for controlling physical access to work areas and hardware (computers, communication devices, etc.) and software media enforcement Inadequate monitoring of the organization premises Inadequate audit logs to detect unauthorized access of the premises Lack of a formal entitlement review process regarding the access rights of the employees in the organization's premises Inadequate maintenance of the records regarding the repairs and modifications of the organization facilities physical components

Asset type: ICT infrastructure

Threat	Vulnerability
Fire	Inadequate physical and environmental security policy and procedures Lack of a uniform physical security policy enforcement Lack of environmental protection Inadequate monitoring of environmental conditions No business continuity plans for recovery of information and information assets Inadequate recovery procedure No formal or informal disaster or recovery plans No concrete assignment of continuity or disaster-related roles and responsibilities Inadequate change management procedure for infrastructure components Improper or inappropriate maintenance of technical facilities Lack of automatic fire suppression system Lack of backup facilities or processes Inadequate backup policy Backup files and systems not available Lack of fire detection devices
Earthquake	Backup files and systems not available Location is in an area susceptible to natural disasters No business continuity plans or procedures for recovery of information and information assets
Flood	Backup files and systems not available Location is in an area susceptible to natural disasters No business continuity plans or procedures for recovery of information and information assets
Storm	Backup files and systems not available Location is in an area susceptible to natural disasters No business continuity plans or procedures for recovery of information and information assets
Tidal surge/wave	Backup files and systems not available Location is in an area susceptible to natural disasters No business continuity plans or procedures for recovery of information and information assets
Contamination	Backup files and systems not available Lack of maintenance of equipment and facilities Location is in an area susceptible to environmental conditions such as contamination, electronic interference extreme temperature, and humidity vermin No business continuity plans or procedures for recovery of information and information assets

Continued

—*Continued*

Threats	Vulnerabilities
Electronic interference	Backup files and systems not available No business continuity plans or procedures for recovery of information and information assets
Extremes of temperature and humidity	Backup files and systems not available Location is in an area susceptible to environmental conditions such as contamination, electronic interference extreme temperature, and humidity vermin No business continuity plans or procedures for recovery of information and information assets Inadequate physical and environmental security policy and procedures Lack of a uniform physical security policy enforcement Lack of environmental protection Inadequate monitoring of environmental conditions Inadequate recovery procedure No formal or informal disaster or recovery plans No concrete assignment of continuity or disaster-related roles and responsibilities Inadequate change management procedure for infrastructure components Improper or inappropriate maintenance of technical facilities Location is in an area susceptible to environmental conditions such as extreme temperature and humidity Lack of backup facilities or processes Inadequate data backup procedure for both software and data Inadequate backup policy
Power fluctuations	Backup files and systems not available No business continuity plans or procedures for recovery of information and information assets Inadequate physical and environmental security policy and procedures Lack of a uniform physical security policy enforcement Lack of environmental protection Inadequate monitoring of environmental conditions Location is in an area susceptible to power fluctuations No uninterruptible power supply equipment No power conditioning equipment Inadequate change management procedure for infrastructure components Improper or inappropriate maintenance of technical facilities

—*Continued*

Threats	Vulnerabilities
Vermin	Backup files and systems not available No business continuity plans or procedures for recovery of information and information assets
Failure of outsourced operations	Backup files and systems not available No business continuity plans or procedures for recovery of information and information assets
Transmission errors	Backup files and systems not available No business continuity plans or procedures for recovery of information and information assets Lack careful planning and laying of cables Lack of properly operation of network equipment Lack of cryptographic means to protect integrity of data
Unauthorized software changes	Backup files and systems not available
Technical failures	Failures in the change management process Improper or inappropriate maintenance of technical facilities Lack of network capacity through improper planning or maintenance Lack of environmental protection Lack of user awareness
Industrial action	Incorrect access rights Lack of audit logs to detect unauthorized use of application Lack of an industrial agreement No incident response and reporting procedures and policies No formally documented procedures for identifying, reporting, and responding to suspected security incidents and violations No concrete assignment of security incidents roles and responsibilities No formal incident review and handling process Inadequate incident handling
Operational staff or user errors	Lack of user awareness Lack of a comprehensive security awareness and training program Lack of means to assess the employee awareness level Inadequate documentation Unskilled staff

Continued

—Continued

Threats	Vulnerabilities
Malicious destruction of data and facilities	Lack of physical security
Sabotage	Lack of physical security
Theft and fraud	Inadequate physical and environmental security policy and procedures
	No documented policies and procedures for physical control of hardware and software
	Inadequate physical security
	Lack of logical access security
	Lack of a uniform policy and procedure for controlling physical access to work areas and hardware (computers, communication devices, etc.) and software media enforcement
	Inadequate monitoring of the organization premises
	Inadequate audit logs to detect unauthorized access of the premises
	Lack of a formal entitlement review process regarding the access rights of the employees in the organization's premises
	Inadequate maintenance of the records regarding the repairs and modifications of the organization facilities physical components
	Inadequate change management procedure for infrastructure components
	No documented and tested security plans for safeguarding the systems and networks
	Lack of a comprehensive security awareness and training program
	Insufficient security training
	No concrete assignment of security roles and responsibilities
	Lack of physical security
Unauthorized data access	Lack of physical security
	Lack of logical access control and audit
Reduced budgets	Inadequate investment in appropriate security controls

Asset type: Software

Threat	Vulnerability
Earthquake	Backup files and systems not available
Fire	Backup files and systems not available
Flood	Backup files and systems not available
	Inadequate data backup procedure for both software and data
Storm	Backup files and systems not available
Tidal surge/wave	Backup files and systems not available
Contamination	Backup files and systems not available
Electronic interference	Backup files and systems not available
Extremes of temperature and humidity	Backup files and systems not available
Power fluctuations	Backup files and systems not available
Vermin	Backup files and systems not available
Failure of outsourced operations	Backup files and systems not available
	Unclear obligations in outsourcing agreements
Transmission errors	Backup files and systems not available
Unauthorized software changes	Backup files and systems not available
	Inadequate reporting and handling of software malfunctions
	Inadequate segregation of duties between software developers and operations staff
	Inadequate supervision of programming staff
	Incorrectly configured or maintained operating system
	Incorrectly configured or maintained security safeguards
	Lack of a firewall
	Lack of configuration management software to enforce configuration management
	Lack of intrusion detection software
	Lack of software configuration lack of software configuration management policies and procedures
	Inadequate reporting and handling of software malfunctions
Use of pirated software	Lack of policy restricting staff to use of licensed software
	Inadequate control of software distribution
	Lack of software auditing
	Unrestricted copying of software
	Inadequate control of software distribution
	Lack of policies in respect of software use
	Uncontrolled copying of data and/or software

Continued

—*Continued*

Threats	Vulnerabilities
Malicious code	Inadequate information security policy No antivirus software Legacy systems Lack of regular update of antivirus software Inadequate education of staff on software viruses Lack of policy for opening email attachments Lack of checks for unauthorized software Lack of policy on using portable storage devices and media before scanning by antivirus software
Web site intrusion	Inadequate software development standards Incorrectly configured or maintained operating system Incorrectly configured or maintained security safeguards Lack of a firewall Lack of intrusion detection software Lack of update of operating system security patches
Software or programming errors	Inadequate system development life cycle procedures Lack of efficient and effective configuration change control Unclear or incomplete specifications Unskilled staff Inadequate engineering code security guidelines for developing web based applications Inadequate security testing of the applications No check for security flaws, covert channels, and back doors as part of the applied software change control procedures Lack of software auditing Inadequate reporting and handling of software malfunctions Inadequate segregation of duties between software developers and operations staff Inadequate supervision of programming staff Inadequate software development standards Inadequate system development life cycle procedures
Sabotage	Incorrect access rights Lack of configuration management controls Lack of logical access security Lack of physical security

—Continued

Threats	Vulnerabilities
Unauthorized data access	Incorrectly configured or maintained operating system Inadequate firewall policies Incorrectly configured or maintained security safeguards Lack of a firewall Lack of intrusion detection software Incorrectly configured or maintained security safeguards Inadequate investment in appropriate security controls
Malicious destruction of data	Incorrectly configured or maintained operating system Incorrectly configured or maintained security safeguards Inadequate investment in appropriate security controls Lack of a firewall Lack of checks for unauthorized software Lack of communication between HR and IT groups in respect of terminated employees leading to such employees still having access to system Lack of intrusion detection software
Theft and fraud	Incorrectly configured or maintained operating system Incorrectly configured or maintained security safeguards Lack of a firewall Lack of checks for unauthorized software Uncontrolled copying of data and or software Lack of a comprehensive security awareness and training program Insufficient security training No concrete assignment of security roles and responsibilities
Social engineering	Lack of awareness of the social engineering threat Lack of policy requiring enquires for information to be withheld until the identity of the requestor can be verified Lack of policy restricting the provision of information by staff over the phone

Continued

—Continued

Threats	Vulnerabilities
Operational staff or user errors	Lack of user awareness Unskilled staff
Technical failures	Lack of user awareness
Denial of service	No antivirus software Incorrectly configured or maintained Lack of a firewall Lack of regular update of antivirus software Inefficient configuration of antivirus software
Reduced budgets	Inadequate investment in appropriate security controls

Asset type: Information and electronic data

Threats	Vulnerabilities
Fire	Inadequate physical and environmental security policy and procedures Lack of a uniform physical security policy enforcement Lack of environmental protection Inadequate monitoring of environmental conditions No business continuity plans for recovery of information and information assets Inadequate recovery procedure No formal or informal disaster recovery plans No concrete assignment of continuity or disaster-related roles and responsibilities Inadequate change management procedure for infrastructure components Improper or inappropriate maintenance of technical facilities Lack of automatic fire suppression system Lack of backup facilities or processes Inadequate backup policy Backup files and systems not available Lack of fire detection devices
Earthquake	Backup files and systems not available Location is in an area susceptible to natural disasters
Flood	Backup files and systems not available Location is in an area susceptible to natural disasters
Storm	Backup files and systems not available Location is in an area susceptible to natural disasters
Tidal surge/wave	Backup files and systems not available Location is in an area susceptible to natural disasters

—Continued

Threats	Vulnerabilities
Contamination	Backup files and systems not available Lack of maintenance of equipment and facilities Location is in an area susceptible to environmental conditions such as contamination, electronic interference extreme temperature, and humidity vermin
Electronic interference	Backup files and systems not available
Extremes of temperature and humidity	Backup files and systems not available Location is in an area susceptible to environmental conditions such as contamination, electronic interference extreme temperature, and humidity vermin Inadequate physical and environmental security policy and procedures Lack of a uniform physical security policy enforcement Lack of environmental protection Inadequate monitoring of environmental conditions Inadequate recovery procedure No formal or informal disaster recovery plans No concrete assignment of continuity or disaster-related roles and responsibilities Inadequate change management procedure for infrastructure components Improper or inappropriate maintenance of technical facilities Location is in an area susceptible to environmental conditions such as extreme temperature and humidity Lack of backup facilities or processes Inadequate data backup procedure for both software and data Inadequate backup policy
Power fluctuations	Backup files and systems not available Inadequate physical and environmental security policy and procedures Lack of a uniform physical security policy enforcement Lack of environmental protection Inadequate monitoring of environmental conditions Location is in an area susceptible to power fluctuations No uninterruptible power supply equipment No power conditioning equipment Inadequate change management procedure for infrastructure components Improper or inappropriate maintenance of technical facilities
Vermin	Backup files and systems not available

Continued

—Continued

Threats	Vulnerabilities
Failure of outsourced operations	Backup files and systems not available Unclear obligations in outsourcing agreements
Transmission errors	Backup files and systems not available Improper or inappropriate cabling Inadequate incident handling
Unauthorized software changes	Backup files and systems not available Easily accessible SCADA devices Inadequate engineering and quality processes for design and code review Inadequate reporting and handling of software malfunctions Inadequate segregation of duties between software developers and operations staff Inadequate supervision of programming staff Incorrectly configured or maintained operating system Incorrectly configured or maintained security safeguards Lack of a firewall Lack of backups Lack of configuration management software to enforce configuration management Lack of intrusion detection software Lack of software configuration management policies and procedures
Operational staff or user errors	Complicated user interface Lack of a comprehensive security awareness and training program Lack of means to assess the employee awareness level Inadequate documentation Unskilled staff Lack of user awareness
Unauthorized dial-in access	Dial-in banner leading to information that can expose the organization to unauthorized dial-in access Lack of an inventory of dial-up lines leading to inability to monitor dial-up access Lack of audit logs to detect unauthorized access Lack of user authentication Lack of intrusion detection software Lack of firewall Lack of policies in respect of dial-up access, modem use, and software use Lack of time restrictions on user access Lack of physical security over telecommunications equipment cabinets Lack of dial back authentication

—*Continued*

Threats	Vulnerabilities
Malicious code	No antivirus software Lack of regular update of antivirus software Inadequate education of staff on software viruses Lack of policy for opening email attachments Lack of control of instant messaging Legacy systems Lack of checks for unauthorized software Lack of policy on using portable storage devices and media before scanning by antivirus software
Web site intrusion	Inadequate firewall policies Inadequate software development standards Incorrectly configured or maintained operating system Lack of intrusion detection software Lack of update of operating system security patches
Unauthorized data access	Inadequate firewall policies Inadequate operating policies for handling, processing, or storing sensitive information Incorrectly configured or maintained application security features Incorrectly configured or maintained operating system Incorrectly configured or maintained security safeguards Lack of a firewall Lack of intrusion detection software Unsecured wireless ports Transmission of unencrypted sensitive data or information Lack of physical security over data communications cabinets Portable devices storing unencrypted data and information Inability to authenticate requests for information No formal policy for the establishment and termination of the access right to information assets Inadequate investment in appropriate security controls Inadequate identity and password policy Unprotected password tables Lack of identification and authentication mechanisms Incorrect access rights Inadequate review of the user's access rights

Continued

—Continued

Threats	Vulnerabilities
Malicious destruction of data	Inadequate firewall policies Inadequate operating policies for handling, processing, or storing sensitive information Incorrectly configured or maintained application security features Incorrectly configured or maintained operating system Incorrectly configured or maintained security safeguards Inadequate investment in appropriate security controls Lack of a firewall Lack of intrusion detection software Unsecured wireless ports Lack of physical security
Theft and fraud	Inadequate firewall policies Inadequate operating policies for handling, processing, or storing sensitive information Incorrectly configured or maintained application security features Incorrectly configured or maintained operating system Incorrectly configured or maintained security safeguards Lack of a firewall Lack of application safeguards leading to fraudulent payments being made Lack of appropriate control of outbound traffic Lack of checks for unauthorized software Lack of safeguards leading to false credentials being created or accepted Lack of effective software change management leading to unauthorized software modifications that could be used to perpetrate a fraud Lack of logical access security Lack of physical security Lack of procedural safeguards leading to fraudulent payments being made Revealing too much information about systems to people without a "need to know" Uncontrolled copying of data and or software Unsecured wireless ports Lack of appropriate control of outbound traffic Lack of a uniform policy and procedure for controlling physical access to work areas and hardware (computers, communication devices, etc.) and software media enforcement Inadequate monitoring of the organization premises

—*Continued*

Threats	Vulnerabilities
	Lack of a formal entitlement review process regarding the access rights of the employees in the organization's premises
	Inadequate change management procedure for infrastructure components
	Lack of a comprehensive security awareness and training program
	Insufficient security training
	No concrete assignment of security roles and responsibilities
Denial of service	Inadequate network management (resilience of routing)
	Incorrectly configured or maintained security safeguards
	Lack of a firewall
	No antivirus software
	Not keeping up to date with security advisories will lead to a known weakness not being corrected in a timely manner
	Lack of regular update of antivirus software
	Inefficient configuration of antivirus software
Software or programming errors	Inadequate system development life cycle procedures
	Unclear or incomplete specifications
	Unskilled staff
Misrouting or rerouting messages	Inadequate user training
	Transmission of unencrypted confidential data
	Lack of proof of receiving a message
Sabotage	Incorrect access rights
	Lack of configuration management controls
	Lack of logical access security
	Lack of physical security
Industrial action	Lack of an industrial agreement
	Incorrect access rights
	Lack of audit logs to detect unauthorized use of application
	Inadequate network administration tools
	Lack of intrusion and prevention systems detection software
	Lack of event management and correlation system
	Lack of data leak systems
	Lack of audit logs to detect unauthorized use of application
	Inadequate audit logs to detect unauthorized access
	Inadequate audit logs to detect malicious use of information systems or applications

Continued

—Continued

Threats	**Vulnerabilities**
Social engineering	Lack of awareness of the social engineering threat Lack of policy requiring enquires for information to be withheld until the identity of the requestor can be verified Lack of policy restricting the provision of information by staff over the phone
Masquerade	Lack of identification and authentication mechanisms Lack of identification of sender and receiver Unprotected password tables Lack of a comprehensive security awareness and training program Inadequate user training Insufficient security training Lack of means to assess the employee awareness level Inadequate identity and password policy
Eavesdropping	Unencrypted communications Lack of physical security over data communications closets or hubs Use of shared ethernet means that all traffic is broadcast to any machine on a local segment Inadequate security controls for the protection of sensitive information being either in storage or during transmission (e.g., data encryption, public key infrastructure, virtual private network technology) Lack of encryption mechanisms
Repudiation	Lack of proof of sending or receiving a message Lack of use of digital signatures
Technical advances such as quantum computing	Inefficient encryption algorithms
Reduced budgets	Inadequate investment in appropriate security controls

APPENDIX C

SUPPLY CHAIN CONTROLS AND VULNERABILITIES

Table C.1 Medusa's security declaration statement

Implementation level factor (security controls)	Yes	No	Partially	Comments
(1) Management of supply chain security				
$R_{1.1}$ Does the organization have a management system that addresses supply chain security? $R_{1.2}$ Does the organization have a person designated as responsible for supply chain security?				
(2) Security plan				
$R_{2.1}$ Does the organization have (a) current security plan(s)? $R_{2.2}$ Does the plan address the organization's security expectations of upstream and downstream business partners? $R_{2.3}$ Does the organization have a crisis management, business continuity, and security recovery plan?				
(3) Asset security				
$R_{3.1}$ Does the organization have in place measures that addresses the physical security of buildings, monitoring and controlling of exterior and interior perimeters, application of access controls that prohibit unauthorized access to facilities, conveyances, loading docks, and cargo areas, and managerial control over the issuance of identification (employee, visitor, vendor, etc.) and other access devices? $R_{3.2}$ Are there operational security technologies that significantly enhance asset protection? For example, intrusion detection or recorded CCTV/DVS cameras that cover areas of importance to the supply chain activity, with the recordings maintained for a long enough period of time to be of use in an incident investigation.				

$R_{3.3}$ Are there protocols in place to contact internal security personnel or external law enforcement in case of security breach? $R_{3.4}$ Are procedures in place to restrict, detect, and report unauthorized access to all cargo and conveyance storage areas? $R_{3.5}$ Are persons delivering or receiving cargo identified before cargo is received or released?				
(4) Personnel security				
$R_{4.1}$ Does the organization have procedures to evaluate the integrity of employees prior to employment and periodically relative to their security duties? $R_{4.2}$ Does the organization conduct specific job appropriate training to assist employees in performing their security duties, for example, maintaining cargo integrity, recognizing potential internal threats to security and protecting access controls? $R_{4.3}$ Does the organization make employees aware of the procedures the company has in place to report suspicious incidents? $R_{4.4}$ Does the access control system incorporate immediate removal of a terminated employee's company-issued identification and access to sensitive areas and information systems?				
(5) Information security				
$R_{5.1}$ Are procedures employed to ensure that all information used for cargo processing, both electronic and manual, is legible, timely, accurate, and protected against alteration, loss, or introduction of erroneous data? $R_{5.2}$ Does an organization shipping or receiving cargo reconcile the cargo with the appropriate shipping documentation?				

Continued

Table C.1 Medusa's security declaration statement—cont'd

Implementation level factor (security controls)	Yes	No	Partially	Comments
$R_{5.3}$ Does the organization ensure that cargo information received from business partners is reported accurately and in a timely manner?				
$R_{5.4}$ Is relevant data protected through use of storage systems not contingent on the operation of the primary data handling system (is there a data backup process in place)?				
$R_{5.5}$ Do all users have a unique identifier (user ID) for their personal and sole use, to ensure that their activities can be traced to them?				
$R_{5.6}$ Is an effective password management system employed to authenticate users, and are users required to change their passwords at least annually?				
$R_{5.7}$ Is there protection against unauthorized access to and misuse of information?				
(6) Goods and conveyance security				
$R_{6.1}$ Are procedures in place to restrict, detect, and report unauthorized access to all shipping, loading dock areas, and closed cargo transport unit storage?				
$R_{6.2}$ Are qualified persons designated to supervise cargo operations?				
$R_{6.3}$ Are procedures in place for notifying appropriate law enforcement in cases where anomalies or illegal activities are detected or suspected by the organization?				
$R_{6.4}$ Are procedures in place to ensure the integrity of the goods/cargo when the goods/cargo are delivered to another organization (transportation provider, consolidation center, intermodal facility, etc.) in the supply chain?				
$R_{6.5}$ Are processes in place to track changes in threat levels along transport routes?				

(7) Closed cargo transport units

(WCO SAFE framework includes a "Seal Integrity Program" described in the Appendix to Annex 1 that sets out procedures regarding the affixing and verification of high-security seals and/or other tamper-detection devices. Personnel filling in this form should review that section of the framework).

$R_{7.1}$ If a closed cargo transport unit is used, are there documented procedures for affixing and recording high-security mechanical seals meeting ISO/PAS 17712 and/or other tamper-detection devices by the party stuffing the cargo unit?				
$R_{7.2}$ If a sealed, closed cargo transport unit is used, are there documented procedures in place to inspect seals for signs of tampering when the custody of conveyances changes during a shipment and to address detected discrepancies?				
$R_{7.3}$ If a closed cargo transport unit is used, is it inspected for contamination by the party stuffing immediately before stuffing?				
$R_{7.4}$ If closed cargo transport units are used, are documented procedures in place for inspecting them immediately before stuffing by the party stuffing them to verify their physical integrity, to include the reliability of the unit-locking mechanisms? A seven-point inspection process is recommended: Front wall Left side Right side Floor Ceiling/roof Inside/outside closure Outside/undercarriage				

Table C.2 Security vulnerabilities and related security controls related with supply chain security

Area	**1. Management of supply chain security**		
Security vulnerability		**Security control**	
$V_{1.1}$	Lack of a supply chain security management system	$C_{1.1}$	Implementation of a supply chain security management system
$V_{1.2}$	Supply chain security role not defined in the organization	$C_{1.2}$	The organization has designated a person responsible for supply chain security
Area	**2. Security plan**		
Security vulnerability		**Security control**	
$V_{2.1}$	Lack of document security plan(s)	$C_{2.1}$	Document security plan(s) exist
$V_{2.2}$	Security plan(s) do not consider the security expectations of upstream and downstream business partners	$C_{2.2}$	Security plan(s) address the organization's security expectations of upstream and downstream business partners
$V_{2.3}$	Lack of crisis management, business continuity, and security recovery plan	$C_{2.3}$	Implementation of crisis management, business continu-ity, and security recovery plan
Area	**3. Asset security**		
Security vulnerability		**Security control**	
$V_{3.1}$	Lack of measures for the physical security of buildings, monitoring and controlling of exterior and interior perimeters, application of access controls that prohibit unauthorized access to facilities, conveyances, load-ing docks, and cargo areas, and managerial control over the issuance of identification (employee, visitor, vendor, etc.) and other access devices	$C_{3.1}$	Implementation of measures that address the physical security of buildings, monitoring and controlling of exterior and interior perimeters, application of access controls that prohibit unauthorized access to facilities, conveyances, loading docks, and cargo areas, and managerial control over the issuance of identification (employee, visitor, vendor, etc.) and other access devices

$V_{3.2}$	Lack of operational security technologies for asset protection (e.g., intrusion detection, recorded CCTV/DVS cameras covering areas of importance to the supply chain activity, with the recordings maintained for a long enough period)	$C_{3.2}$	Implementation of operational security technologies for asset protection (e.g., intrusion detection, recorded CCTV/DVS cameras covering areas of importance to the supply chain activity, with the recordings maintained for a long enough period to be of use in an incident investigation)
$V_{3.3}$	Lack of protocols to contact internal security personnel or external law enforcement in case of security breach	$C_{3.3}$	Protocols are in place to contact internal security personnel or external law enforcement in case of security breach
$V_{3.4}$	Lack of procedures that restrict, detect, and report unauthorized access to all cargo and conveyance storage areas	$C_{3.4}$	Implementation of procedures that restrict, detect, and report unauthorized access to all cargo and conveyance storage areas
$V_{3.5}$	Lack of procedures to identify persons delivering or receiving cargo, before cargo is received or released	$C_{3.5}$	Implementation of procedures to identify persons delivering or receiving cargo, before cargo is received or released
Area	**4. Personnel security**		
Security vulnerability		**Security control**	
$V_{4.1}$	Lack of procedures to evaluate the integrity of employees prior to employment and periodically, relative to their security duties	$C_{4.1}$	Implementation of procedures to evaluate the integrity of employees prior to employment and periodically, relative to their security duties
$V_{4.2}$	Lack of specific, job appropriate training, to assist employees in performing their security duties (e.g., maintaining cargo integrity, recognizing potential internal threats to security and protecting access controls)	$C_{4.2}$	Implementation of specific, job appropriate training, to assist employees in performing their security duties (e.g., maintaining cargo integrity, recognizing potential internal threats to security and protecting access controls)

Continued

Table C.2 Security vulnerabilities and related security controls related with supply chain security—cont'd

$V_{4.3}$	Lack of measures for the dissemination to the employees of the installed procedures for reporting suspicious incidents	$C_{4.3}$	Implementation of procedures to make employees aware of the implemented procedures for reporting suspicious incidents
$V_{4.4}$	Lack of procedures that restrict, detect, and report unauthorized access to all cargo and conveyance storage areas	$C_{4.4}$	Implementation of procedures to restrict, detect, and report unauthorized access to all cargo and conveyance storage areas
Area	**5. Information security**		
Security vulnerability		**Security control**	
$V_{5.1}$	Lack of procedures to ensure that all information used for cargo processing, both electronic and manual, is legible, timely, accurate, and protected against alteration, loss, or introduction of erroneous data	$C_{5.1}$	Implementation of procedures to ensure that all information used for cargo processing, both electronic and manual, is legible, timely, accurate, and protected against alteration, loss, or introduction of erroneous data
$V_{5.2}$	Lack of appropriate shipping documentation reconciling the cargo, when an organization is shipping or receiving cargo	$C_{5.2}$	Implementation of procedures to ensure that an organization shipping or receiving cargo reconciles the cargo with the appropriate shipping documentation
$V_{5.3}$	Lack of procedures to ensure that cargo information received from business partners is reported accurately and in a timely manner	$C_{5.3}$	Implementation of procedures to ensure that cargo information received from business partners is reported accurately and in a timely manner
$V_{5.4}$	Lack of a data backup process and storage system(s) for relevant data	$C_{5.4}$	Implementation of a data backup process and storage system(s) for relevant data
$V_{5.5}$	Users are not required to have unique identifiers for their personal and sole use	$C_{5.5}$	Implementation of procedures to ensure that all users have a unique identifier (user ID) for their personal and sole use that can be used to trace their activities

$V_{5.6}$	Lack of an effective password management system to authenticate users and to ensure at least annual password change	$C_{5.6}$	Implementation of an effective password management system to authenticate users and to ensure that the users change their passwords at least annually
$V_{5.7}$	Lack of protection against unauthorized access to and misuse of information	$C_{5.7}$	Implementation of procedures and/or technical controls to ensure protection against unauthorized access to and misuse of information
Area	**6. Goods and conveyance security**		
Security vulnerability		**Security control**	
$V_{6.1}$	Lack of procedures to restrict, detect, and report unauthorized access to all shipping, loading dock areas, and closed cargo transport unit storage	$C_{6.1}$	Implementation of procedures to restrict, detect, and report unauthorized access to all shipping, loading dock areas, and closed cargo transport unit storage
$V_{6.2}$	Lack of procedures to ensure that qualified persons are designated to supervise cargo operations	$C_{6.2}$	Implementation of procedures to ensure that qualified persons are designated to supervise cargo operations
$V_{6.3}$	Lack of procedures for notifying appropriate law enforcement in cases where anomalies or illegal activities are detected or suspected by the organization	$C_{6.3}$	Implementation of procedures for notifying appropriate law enforcement in cases where anomalies or illegal activities are detected or suspected by the organization
$V_{6.4}$	Lack of procedures to ensure the integrity of the goods/cargo when the goods/cargo are delivered to another organization (transportation provider, consolidation center, intermodal facility, etc.) in the supply chain	$C_{6.4}$	Implementation of procedures to ensure the integrity of the goods/cargo when the goods/cargo are delivered to another organization (transportation provider, consolidation center, intermodal facility, etc.) in the supply chain
$V_{6.5}$	Lack of procedures to track changes in threat levels along transport routes	$C_{6.5}$	Implementation of procedures to track changes in threat levels along transport routes

Continued

Table C.2 Security vulnerabilities and related security controls related with supply chain security—cont'd

Area	7. Closed cargo transport units security		
Security vulnerability		**Security control**	
$V_{7.1}$	Lack of documented procedures, if a closed cargo transport unit is used, for affixing and recording high-security mechanical seals meeting ISO/PAS 17712 and/or other tamper-detection devices by the party stuffing the cargo unit	$C_{7.1}$	Implementation of documented procedures, if a closed cargo transport unit is used, for affixing and recording high-security mechanical seals meeting ISO/PAS 17712 and/or other tamper-detection devices by the party stuffing the cargo unit
$V_{7.2}$	Lack of documented procedures, if a sealed, closed cargo transport unit is used, to inspect seals for signs of tampering when the custody of conveyances changes during a shipment and to address detected discrepancies	$C_{7.2}$	Implementation of documented procedures, if a sealed, closed cargo transport unit is used, to inspect seals for signs of tampering when the custody of conveyances changes during a shipment and to address detected discrepancies
$V_{7.3}$	Lack of procedures, if a closed cargo transport unit is used, to ensure inspection for contamination by the party stuffing immediately before stuffing	$C_{7.3}$	Implementation of procedures, if a closed cargo transport unit is used, to ensure inspection for contamination by the party stuffing immediately before stuffing
$V_{7.4}$	Lack of documented procedures, if closed cargo transport units are used, to ensure their inspection immediately before stuffing by the party stuffing them, to verify their physical integrity, and to ensure the reliability of the unit-locking mechanisms	$C_{7.4}$	Implementation of documented procedures, if closed cargo transport units are used, to ensure their inspection immediately before stuffing by the party stuffing them, to verify their physical integrity, and to ensure the reliability of the unit-locking mechanisms. A seven-point inspection process is recommended: (1) front wall, (2) left side, (3) right side, (4) floor, (5) ceiling/roof, (6) inside/outside closure, and (7) outside/undercarriage

INDEX

Note: 'Page numbers followed by "f" indicate figures and "t" indicate tables.'

A

Active vessel traffic management and information system (A-VTMIS), 29
Administration module, 55
Advanced security intelligence engine
  asset identification component, 64
    risk assessment service, 64–65
  components, 64
  evaluation component, 64–65
  functions, 65
  holistic approach, 66
AIS navigation tracking system, 3–4
Annual loss expectancy (ALE), 34
Automated border control (ABC) systems, 128–129
Autonomous Maritime Surveillance System (AMASS) project, 30
Autonomous/semiautonomous mechanical physical systems, 3
Awareness rising campaigns, 133

B

Business partners association, 108

C

Cargo management services, 13
Cargo ships, 3
Certificate policies (CPs), 131
Certification authorities (CAs), 130–131
Cloud computing technology, 132
Commercial seaports, 1
Common Attack Pattern Enumeration and Classification (CAPEC), 105
  version usage, 106
Community collaboration suite, 54–55
Community e-Library, 55
Complex physical–cyber systems, 3
Computer Emergency Response Team–European Union (CERT-EU), 10–11
Computer Security Incident Response Team, 5–6
Consequence scale, 72, 78t
cPPP Initiative, 32
CRAMM, 37–38
Critical information infrastructure protection (CIIP)
  applications availability, 46
  assessment of, 50, 50t
  CI sectors, 46
  criteria, 46
  Department of Homeland Security and the Department of Energy, 50–51
  description of, 46, 47t–49t
  emergency response strategies, 50–51
  policies and regulations implementation, 51
  risk reduction, 51
  safety threats, 50–51
  stages, 50
  standards and methodologies, 46
Critical information infrastructure protection (CIIP) Directive, 31
Critical information infrastructures (CIIs), 6–7, 125
  commercial ports, 7
  critical information infrastructure protection (CIIP). *See* Critical information infrastructure protection (CIIP)
  cyber/physical security management (CYSM) risk assessment tool, 55f. *See also* Cyber/physical security management (CYSM) risk assessment tool
  cybersecurity regulations and standards
    CIIP Directive, 31
    cPPP Initiative, 32
    Cybersecurity Strategy for the European Union, 31
    eIDAS Regulation, 31

Critical information infrastructures (CIIs) (*Continued*)
Enhanced Privacy Directive, 32
European Agenda on Security, 31
European Parliament, 31
ICT security management standards, 32
International Ship and Port Facility Security (ISPS) Code, 31
ISO/IEC 27001 36, 32
ISO/IEC 27002:2005, 33–34
ISO/IEC 27005:2008 38, 33
NIS Directive, 31
USA H.R. 3878, 32
international standards, 27
maritime ecosystem, 27–28
physical risk assessment, 27–28
risk analysis and evaluation, 27–28
risk management, 27
safety management. *See* Safety management
security management, 27. *See also* Security management
Critical infrastructure (CI), 3
commercial ports, 7–8, 8f
complex business processes, 7
critical e-port/supply chain services, 7
cybersecurity, 7
definition, 7
information and communication technology (ICT) system, 7
port layers, 8–9
maritime security organizations. *See* Maritime security organizations
port services
cargo management services, 13
customs clearance approval, 17–19
data historian database, 20
electronic and mobile (e/m) port service, 14–15. *See also* Electronic and mobile (e/m) port service
Electronic Data Interchange system and reports, 17–19
four-level framework, 14
goal and functions, 13
human machine interface (HMI), 20
inland logistics services, 14
Integrated Customs Information System (ICIS), 17
integration services, 14
"MarineTraffic" web platform, 17–19
multichannel communication, 14
online sophistication maturity level, 14
Port Community System (PCS), 17–19
programming logic controllers (PLCs), 20
remote terminal unit (RTU) devices, 20
SCADA master terminal unit (MTU), 20
security critical, 15
Supervisory Control and Data Acquisition (SCADA) systems, 20
supervisory stations, 20
supply chains (SCs), 16
supply chain services (SCSs), 16, 20–22, 21t–22t. *See also* Supply chain services (SCSs)
trailer cranes and forklifts, 20
vessel arrival and vehicles uploading phase, 17, 19f
vessel management services, 13
safety (physical security), 9, 10f
secure (cybersecure), 9, 10f
Cumulative impact level (CIL), 119
Cumulative vulnerability assessment, 112–116
Cumulative vulnerability level (CVL), 116
Customs clearance approval, 17–19
CVE date usage, 105
CVE name support styles, 105
Cyberattacks, 3–5
Cyber/physical security management (CYSM) risk assessment tool, 55f
administration module, 56, 56f
architecture
advanced security intelligence engine. *See* Advanced security intelligence engine
CYSM system's workflow engine, 63
execution engine component, 63
integrated and interconnected subsystems, 60, 61f

interaction component, 63–64
semantic modeling component, 62–63
web interactive component, 60–62
automation, 52
broad threat analysis, 52
collaboration, 52–53
collaborative, multiattribute, group decision-making algorithm, 54
conformance approach, 53
customization phase, 56–57
CYSM Security Assessment Services, 54
decision-making, 53
descriptive approach, 53
EU CIPS Project, 53
evaluation phase, 57–59, 58f
functions and procedures, 53
holistic, 52
implementation, 52
interdependencies, 52
interdependent threats, 51
knowledge codification, 52
personalization, 52
requirements, 52
risk assessment initiation phase, 57, 58f
risk assessment results calculation and review phase, 59, 59f
role-specific content views, 53
security avoidance, 52
security policy–growing mechanism, 54
security policy reporting phase, 59–60, 60f
service-oriented, 53
social, collaborative working environment, 54
standards compatibility, 52
system components, 54–56
community portal, 54–55
port private portal, 55
time and resource economical, 52
user centric, 52
Cyber risk assessment, 153, 153t
report, 153
Cybersecurity, 2, 7
ENISA, 3
regulations
CIIP Directive, 31
cPPP Initiative, 32
Cybersecurity Strategy for the European Union, 31
eIDAS Regulation, 31
Enhanced Privacy Directive, 32
European Agenda on Security, 31
European Parliament, 31
ICT security management standards, 32
ISPS code, 31
NIS Directive, 31
USA H.R. 3878, 32
risk management, 10
standards
ISO/IEC 27001 36, 32
ISO/IEC 27002:2005, 33–34
ISO/IEC 27005:2008 38, 33
Cybersecurity Strategy for the European Union, 31

D

Department of Homeland Security and the Department of Energy, 50–51

E

EBIOS, 38–39
EC CIPS project, 79
Economic crisis, 130
eIDAS Regulation, 31
Electronic and mobile (e/m) port service
high business impact, 16
maritime entities, 16
low business impact, 15
medium business impact, 15
maritime entities, 16
maritime environment support, 15
security levels, 15
Electronic Data Interchange system and reports, 17–19
Electronic solutions, 3
Enhanced business declaration, 110
Enhanced Privacy Directive, 32
Enhanced Security Declaration, 107
ENISA, 3, 36
Entry's chain impact level (ECIL), 119
Entry's chain vulnerability level (ECVL), 115
EU Directive 2010/65/EU23, 126

EU FP7 Security Research programs, 29–30
EU project CORE, 105
EU project MITIGATE, 106
European Agenda on Security, 31
European Maritime Safety Agency (EMSA), 11
European Network and Information Security Agency (ENISA), 13
European Parliament, 31
European Sea Ports Organisation (ESPO), 11–12
EU Seventh Framework Programme, 29

F

FLAGSHIP, 29
FRONTEX requirements, 129

G

Global trade and economy, 3

H

Human machine interface (HMI), 20

I

ICT security management standards, 32
ILO Port Health and Safety systems, 28
Individual chain impact level (ICIL), 119
Individual chain vulnerability level (ICVL), 113–115, 114t–115t
Individual propagated chain impact (IPCI), 119
Individual vulnerability assessment, 112, 114t
Industrial Control Systems (ICSs), 4–5
Information and communication technology (ICT) system, 3, 7
  infrastructure
    data, 145
    physical infrastructure, 145
    services and applications, 145
    software, 145
    telco infrastructure equipment, 145
    users, 145
  port layers
    information and electronic data, 9
    network, 9
    physical, 8
    services, 9
    systems and software, 9
    users/processes, 9
Information security management system (ISMS), 32
Information sharing, 133
Inland logistics services, 14
Inland waterways, 12
Innovation and Networks Executive Agency (INEA), 12
Integrated Customs Information System (ICIS), 17
Integration services, 14
Intelligence building, 133
International Convention for the Prevention of Pollution from Ships (MARPOL), 10
International Convention for the Safety of Life at Sea (SOLAS), 10
International Finance Corporation, 28–29
International Maritime Organization (IMO), 10
International merchant marine community, 12
International Port Community Systems Association, 5
International Ship and Port Facility Security (ISPS) Code, 10, 31
InterOPERAble Approach to European Union MARitime Security Management (OPERAMAR) project, 30
INTERPOL, 12
ISAMM, 41
ISO 28000:2007, 103
ISO 28001:2007, 103
ISO/IEC 27001 36, 32
ISO/IEC 27002:2005, 33–34
ISO/IEC 27005:2008, 104
ISO/IEC 27005:2008 38, 33
IT-Grundschutz, 39

L

Legal frameworks, 1–2
Likelihood assessment, 80
  cascading risk assessment, 84, 85t, 86

consequence analysis, 82–83
partial and overall SC risk assessment evaluation, 83–87
risk assessment, 83
risk mitigation, 86
threat and vulnerability values, 82
Likelihood scale, 72, 77t
LNG storage facilities, 4–5
Local port cloud, 132
Logistic software systems, 3

M

MAGERIT, 40
Management module, 55, 57, 58f
Management systems, 1–2
MarineTraffic Internet Automatic Identification System (AIS), 17–19
Maritime community cloud, 132
Maritime cyber education and cyber skills development, 133
Maritime cybersecurity management, 5
Maritime cybersecurity market, 134
Maritime logistics, 3
and supply chains, 3–4, 4f
Maritime Safety and Environmental portal, 29
Maritime scientific community, 6
Maritime security awareness, 126
Maritime Security Council, 12–13
Maritime security organizations, 9
European Maritime Safety Agency (EMSA), 11
European Network and Information Security Agency (ENISA), 13
European Sea Ports Organisation (ESPO), 11–12
Innovation and Networks Executive Agency (INEA), 12
International Maritime Organization (IMO), 10
Maritime Security Council, 12–13
NATO's Alliance Maritime Strategy, 10–11
Maritime Security Risk Analysis Model (MSRAM), 28
Maritime-specific and supply chain cyber risks, 134
Maritime supply chain risk assessment
asset level
accuracy, 105
CAPEC version usage, 106
chain length, 116
Common Attack Pattern Enumeration and Classification (CAPEC), 105
cumulative impact level (CIL), 119
cumulative vulnerability level (CVL), 116
CVE date usage, 105
CVE name support styles, 105
documentation, 105–106
entry points identification, 116
entry's chain impact level (ECIL), 119
entry's chain vulnerability level (ECVL), 115
EU project CORE, 105
EU project MITIGATE, 106
high-level requirements, 104–105
identify attacker profile, 116
impact analysis, 117–119
individual chain impact level (ICIL), 119
individual chain vulnerability level (ICVL), 113–115, 114t–115t
individual propagated chain impact (IPCI), 119
intentional and unintentional compromise, 103
ISO 28000:2007, 103
ISO 28001:2007, 103
ISO/IEC 27005:2008, 104
isolated threats and vulnerabilities, 103
Medusa methodology, 104
MITIGATE. *See* MITIGATE
propagated impact level (PIL), 119
propagated metrics measurement, 116–117
real-time risk assessment, 104
requirements and recommendations, 104
risk assessment, 120
risk management method, 104
risk mitigation, security controls selection, 121, 122f–123f
supply chain service (SCS), 103
vulnerability chains, 116

Maritime supply chain risk assessment (Continued)
entity level
analysis, 80
auditable, 80
business-centric, 79
collaborative, 79
EC CIPS project, 79
federal information systems, 68
holistic view, 79
implementable, 79
ISO 28001, 67–68
likelihood assessment, 80. *See also* Likelihood assessment
Medusa system. *See* Medusa system
risk management international standards, 67
sector-independent, 80
standards compliance, 79
supply chain graph models. *See* Supply chain graph models
supply chain (SC) security, 67–68
threat assessment, 81
threat scenario identification, 80
validation scenarios. *See* Validation scenarios
vulnerability assessment, 81–82
Maritime surveillance systems, 30
Medusa's security declaration statement, 176t–179t
Medusa system
components, 87, 87f
Medusa technological infrastructure, 89, 89f–91f
SCS risk assessment components, 88
SCS risk assessment information assets, 88
system users, 88
"vehicles' transport" supply chain service, 91–92, 92f
MEHARI, 40–41, 44
MITIGATE
business processes, 107
Enhanced Security Declaration, 107
objectives, 106
structure
assets interdependencies modeling, 110
business partners association, 108
enhanced business declaration, 110
mapping, 108, 109t
SCS analysis, 108–110
SCS cyber assets identification, 108
SCS cyber threat analysis, 110
SCS's business processes (SCSBPs) identification, 108
standard RA blocks, 107
supply chain risk assessment scope, 108
vulnerability analysis, 111–117
MSRAM-PLUS/FORETELL, 28
Multichannel communication, 14

N

NATO Computer Incident Response Capability (NCIRC), 10–11
NATO Maritime Interdiction Operational Training Center, 10–11
NATO's Alliance Maritime Strategy, 10–11
Navigation systems, 3–4
Next generation maritime research, 134
NIS Directive, 31
NIST 800-30, 36–37

O

OCTAVE, 37

P

Physical attacks, 4–5
Physical risk assessment, 152, 152t
report, 152, 152t
Physical security, 2
Plan-Do-Check-Act cycles, 32
Port collaboration suite, 55
Port community system (PCS), 4–5, 17–19, 125
Port e-Library, 55
Port state control, 2
Preparatory Action on Security Research, 29–30
Private cloud, 132
Programming logic controllers (PLCs), 20
Propagated impact level (PIL), 119
Public key infrastructures (PKIs), 130

R

RAND Corporation, 4–5
Regional port cloud, 132
Remote terminal unit (RTU) devices, 20
Risk assessment methodology, 68
Risk Assessment module, 55, 58f
  assets identification, 58
  control identification, 59
  impact assessment, 58
  threat assessment, 59
Risk Assessment Results module, 55, 59, 59f
Risk management international standards, 67
Risk scale, 72, 79t
River Information Services, 12

S

SafePort, 29
Safety management
  critical maritime infrastructure protection, 31
  EU FP7 Security Research programs, 29–30
  FLAGSHIP, 29
  ILO Port Health and Safety systems, 28
  International Finance Corporation, 28–29
  interoperability, ports' CI systems, 30
  maritime risk assessment systems, 28
  Maritime Security Risk Analysis Model (MSRAM), 28
  maritime surveillance systems, 30
  MSRAM-PLUS/FORETELL, 28
  physical threats, 29
  Preparatory Action on Security Research, 29–30
  SafePort, 29
  Safety4Sea, 29
  World Bank Group, 28–29
Safety4Sea, 29
SCADA master terminal unit (MTU), 20
SCS's business processes (SCSBPs) identification, 108
Sea Border Surveillance (SEABILLA) project, 30
Seaports, 12
SECure CONtainer Data Device (SECCONDD) project, 30–31
Security awareness, CYSM questionnaire, ports and port operators
  confidential response, 146
  critical ICT infrastructure, 145
  information, 145
  security plan, 146
  security practices, 150
  vulnerability, 145
Security Declaration Statement (SDS), 70, 70t
Security incident management handling, 127–128
Security management
  annual loss expectancy (ALE), 34
  assessment criteria, 35
  asset identification, 35
  automated tools, 36
  collaboration capabilities, 36
  collaborative capabilities, 45
  computational capabilities, 36
  context establishment, 35
  cost, 36
  CRAMM, 37–38
  disadvantages, 44
  EBIOS, 38–39
  ENISA repository, 36
  evaluation scale, 36
  impact evaluation, 36
  incident, asset, 35
  ISAMM, 41
  IT-Grundschutz, 39
  MAGERIT, 40
  maritime external entities, 44
  MEHARI, 40–41, 44
  methods and criteria, 41, 43t
  NIST 800-30, 36–37
  OCTAVE, 37
  physical and cyber layers and assets, 34
  quantitative data, 41–44
  RA/RM support, 36
  required skills, 36
  risk analysis, 35
  risk analysis solutions, 45–46
  risk assessment, 35
  risk communication, 35

Security management *(Continued)*
risk evaluation, 36
risk monitoring and review, 35
risks, 34–35
risk treatment, 35
scope, 35
security-related legislation, 34
standards compliant, 36
target group, 35
threat analysis, 35
threat, asset, 34
threat, asset, needs, 34
threat, critical asset, 35
Security plan
issues, 149–150, 149t–150t
management, 146–147, 146t
identification and analysis, 147, 147t
model, 146, 146t
perception, 146, 146t
schemes, 146, 146t
regulations/legislation selection, 148–149, 148t–149t
Security Policy Reporting module, 55, 59–60, 60f
Security practices
activities, 151, 151t
port status, 155, 155t
cyber risk assessment, 153, 153t
cyber security risks and incidents report, 153
enterprise level, 150, 150t
enterprises collaboration, 152, 152t
issues, 154, 154t
issues, approach/solution, 151, 151t
organization training, 154, 154t
physical risk assessment, 152, 152t
physical security risks and incidents report, 152, 152t
standards, 154–155, 154t–155t
Security System for Maritime Infrastructure, Ports, and Coastal Zones (SECTRONIC) project, 31
Security Upgrade for Ports (SUPPORT) project, 31
Software-based satellite communications, 20
SOLAS Convention, 1–2
port facilities, 2
port state control, 2
safety concept, 1–2
standards, 2
tacit acceptance procedure, 2
Supervisory Control and Data Acquisition (SCADA) systems, 4–5
elements of, 23
human machine interface (HMI) system, 24–25
port services, 20
vehicles robbery attack, 24–25, 24f
vulnerabilities, BPMN events, 25
Supply chain graph models
consequence scale, 72, 78t
likelihood scale, 72, 77t
risk assessment methodology, 68
risk scale, 72, 79t
SC-directed dependency graphs, 68, 69f
Security Declaration Statement (SDS), 70, 70t
security requirements, 70
threat categorization, 72, 73t–74t
threat scale, 72, 76t
vulnerabilities and controls, 70, 71t, 72, 75t
Supply chain intelligence, 3–4
Supply chain risk assessment (SCRA) methodology, 106
Supply chains (SCs), 3, 16
and maritime logistics, 3–4, 4f
security vulnerabilities and controls, 180t–184t
Supply chain services (SCSs), 7, 103
vehicles transport chain, 16–17, 17f, 18t
remote access, 22–23
SCADA attack, 23–25, 24f–25f
threat analysis, 20–22, 21t–22t
Surveillance of Borders, Coastlines and Harbors project, 30
Synergetic international response, 130

T

Tacit acceptance procedure, 2
Threat analysis
ICT infrastructure, 161t–164t
information and electronic data, 168t–174t
physical infrastructure, 157t–160t
software, 165t–168t

Threat scale, 72, 76t
Traffic management systems, 29
Trans-European network of Motorways of the Sea, 12
Trans-European Transport Network (TEN-T), 12

U

Underwater Coastal Sea Surveyor project, 30
USA Congressional Research Service, 4–5
USA H.R. 3878, 32
US government agencies, 12

V

Validation scenarios
- cascading dependency risks validation, 98–102, 100t–101t
- expected consequences variation, 94–96, 96t
- security controls implementation, 92–94, 95t
- threat scenarios, occurrence probability, 96–98, 98t
- Vehicles Transport Service, 92

Vessel management services, 13
Vulnerability analysis, MITIGATE
- confirmed individual vulnerabilities
  - confirmed vulnerabilities, 111
- cumulative vulnerability assessment, 112–116
- individual vulnerability assessment, 112, 114t
- potential/unknown identification, 111

Vulnerability assessment, 81–82

W

Web interactive component
- application function, 61
- Control Panel, 62
- Dockbar, 62
- information and content, 60
- Manage menu, 62
- portlet applications, 61
- Toggle Edit Controls, 62

World Bank Group, 28–29

FSC
www.fsc.org
MIX
Papier | Fördert gute Waldnutzung
FSC® C083411

Kolibri 360 GmbH
Ferdinand-Jühlke-Straße 7
99095 Erfurt, Deutschland
produktsicherheit@kolibri360.de